ADHESION 11

This volume is based on papers presented at the 24th annual conference on Adhesion and Adhesives held at The City University, London

Previous conferences have been published under the titles of
Adhesion 1–10

ADHESION 11

Edited by

K. W. ALLEN

Adhesion Science Group,
The City University, London, UK

ELSEVIER APPLIED SCIENCE
LONDON and NEW YORK

ELSEVIER APPLIED SCIENCE PUBLISHERS LTD
Crown House, Linton Road, Barking, Essex IG11 8JU, England

Sole Distributor in the USA and Canada
ELSEVIER SCIENCE PUBLISHING CO., INC.
52 Vanderbilt Avenue, New York, NY 10017, USA

WITH 47 TABLES AND 100 ILLUSTRATIONS

ⓒ ELSEVIER APPLIED SCIENCE PUBLISHERS LTD 1987
Softcover reprint of the hardcover 1st edition 1987
ⓒ CONTROLLER, HMSO, LONDON 1987—Chapter 1

British Library Cataloguing in Publication Data
Conference on Adhesion and Adhesives *(24th:*
1986: City University, London)
Adhesion 11
I. Title II. Allen, K. W.
541.3′453   QD506
ISBN-13: 978-94-010-8036-1   e-ISBN-13: 978-94-009-3433-7
DOI: 10.1007/978-94-009-3433-7

Library of Congress CIP data applied for

The selection and presentation of material and the opinions expressed are the sole responsibility of the author(s) concerned.

Special regulations for readers in the USA

Preface

Each year, about Easter time, we are reminded of the continuing growth in the use of adhesives. The Annual Conference here at The City University provides a focal point for discussion, both formal and informal, on expansion and development. For those who attend the Conference its rewards and usefulness are immediate and obvious, but its influence extends to a wider audience through the regular publication of the papers. Here in Volume 11 of the series we offer the latest instalment, the papers presented and discussed in 1986.

May I express gratitude to all those who, by their continuing support, both in presenting papers and in attending the meeting, make the whole event so successful and rewarding. Thank you all, very much.

K. W. ALLEN

Contents

Chapter 1

THE STRENGTH AND DURABILITY OF ADHESIVE JOINTS MADE UNDERWATER

M R BOWDITCH, J D CLARKE and K J STANNARD

Admiralty Research Establishment

INTRODUCTION AND BACKGROUND

1.   The bulk of the work to be described was undertaken on a
repayment basis for the Department of Energy who were
interested in the feasibility of using adhesively assisted
repairs to steel offshore structures in the North Sea. The
work was carried out by what is now the Admiralty Research
Establishment at Holton Heath in collaboration with service
colleagues at Dunfermline and with industry. The Holton Heath
responsibility was for the development of suitable adhesive
formulations and for the investigation of compatible surface
preparation and pretreatment routes. It is this latter work
which forms the basis of this paper.

2.   Because of Naval interest in such materials which may be
applied underwater, investigations had been undertaken at
Holton Heath prior to the commencement of this work and during
these early investigations the decision was made to aim for
inherently water resisting materials, and to avoid the water
scavenging approach. Much of the work which had already been
done prior to our involvement had involved the use of water
scavenging components to achieve the displacement of traces of
residual water from the steel adherend after bulk water had
been mechanically displaced. Our view was that materials with
hydrophilic qualities in the uncured condition would probably
continue to demonstrate such properties after cure, thus
prejudicing chances of long term survival in a situation where
it was energetically favourable for water rather than adhesive
to occupy the highly active steel surface. Our objective
therefore was the development of resin formulations which were
essentially hydrophobic and to attempt in some way to create
the favourable conditions underwater which would parallel those
found in the atmosphere, with adhesive tending to spread
spontaneously over the surface of the steel substrate.

3.   It was also recognised that, in order to obtain the high
modulus, low shrinkage product necessary for structural load
bearing applications, quite highly filled systems would have to
be used and the rheology of the adhesives in the uncured
condition would be of great importance. A compromise in terms
of viscosity had to be arrived at. Too high a viscosity would
lead to a situation where effective 'wetting' of the relatively
rough substrate surface would not be achieved during the
timescale of the application whilst an otherwise ideal, low

viscosity would be incompatible with the essential qualities
indicated above and, indeed, with the practicalities of
application in many cases.

4.   In addition, any useful material would be relatively
insensitive to minor departures from compositional
stoichiometry and would have a pot life such that it could be
used by a diver.   Such an adhesive would also be capable of
curing at temperatures down to about 5°C in thin section and
yet, at the same time, it would cure without cracking,
splitting or the formation of significant voids in thick
section (~150 mm).   All these requirements placed further
constraints on formulation.

THE ADHESIVE AND ITS BULK PROPERTIES

5.   It was against this background that three epoxide based
adhesive formulations emerged as candidates for more detailed
evaluation.   Ultimately, the product UW45 was chosen as the
preferred composition[1] and Table 1 gives details of some of
its bulk properties.

TABLE 1

SOME BULK PROPERTIES OF THE ADHESIVE UW45

PROPERTY	VALUE
Tensile Strength, MPa	16.9
Compressive Strength, MPa	28.8
Shear Strength, MPa	24.1
Young's Modulus, MPa	2500
Compressive Modulus, MPa	1966
Bulk Modulus, GPa	5.3
Poisson's Ratio, μ	0.44
Shrinkage on Cure, %	1.97

6.   The viscosity of UW45 is shown in Figure 1 at two shear
rates and it can be seen that the material shear-thins with a
viscosity reduction of about 50% associated with a tenfold
increase in shear rate.   A very large increase in viscosity can
be seen to be associated with a quite small change in
temperature in the range relevant to North Sea usage.
Nevertheless, it has been shown that the adhesive can be
applied manually at temperatures down to ~3°C to make strong
joints to steel which are indistinguishable from others made at
+20°C at the same time.

3

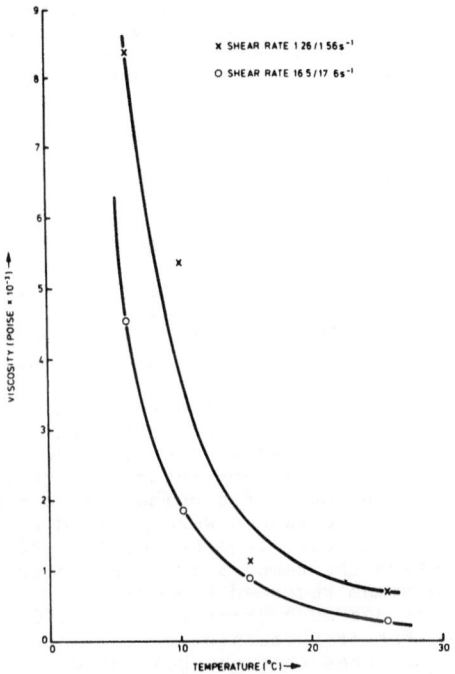

FIG 1 VISCOSITY/TEMPERATURE RELATIONSHIP OF UW45

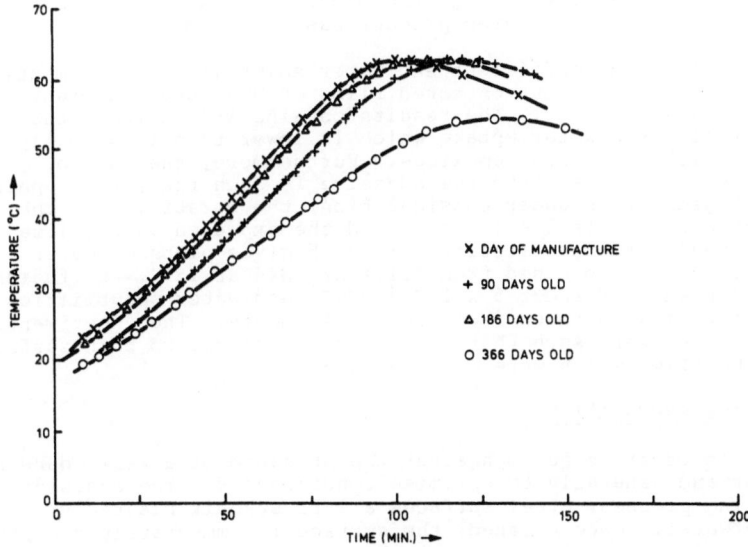

FIG.2.THE EFFECT OF AGEING ON THE REACTIVITY OF UW45 AS INDICATED BY POT LIFE MEASUREMENT

4

7.	The pot-life of the adhesive is reflected in the curves presented in Figure 2.	The results were obtained by measuring the temperature at the centre of a cube of a mix of 75 mm side at recorded intervals.	The time to develop peak temperature and its magnitude is a measure of the reactivity of the resin system.	It can be seen that there is no important reduction in the reactivity of the system after a period of 6 months, suggesting that the shelf life for the material is at least 6 months under standard laboratory storage conditions.

8.	Figure 3 indicates the time required to develop maximum adhesion at 30°C and at 6°C; the latter temperature being taken as typical of that to be found at depth in the North Sea.	It can be seen that, at 6°C, 50% of the maximum adhesion is attained after a period of about 3 days whereas at 30°C a comparable state of cure is achieved after only about 18 hours.

9.	The behaviour of the adhesive in thick section was assessed by casting a block of the material measuring 305 mm x 305 mm x 152 mm in a mild steel container surrounded by a large bulk of water at 8°C.	The temperature at the vertical centre of the block was monitored at a number of points over its cross-section and these values were plotted against time. Figure 4 gives the results together with a sketch indicating the points at which the temperature was monitored.	It can be seen that the maximum recorded temperature, at the very centre of the block, was not much greater than 100°C.	The temperature at point 6 was that of the outside of the steel container and its stability indicates that the bulk of water was such that its temperature was not significantly influenced by the heat generated within the bulk of the adhesive.	After cure, the resin block was cut through the middle for examination and an even-textured, well cured product was revealed.

10.	Diffusion coefficient and water solubility measurements were also carried out on cured films of UW45 under ambient pressure conditions.	The results obtained[2] indicate that the material has a water uptake which is lower than is normally found with cold cured epoxides.	Furthermore, the rate of diffusion of water into the adhesive is such that, for repair joint geometries under consideration, the durability of the jointed repair is likely to exceed the expected useful life of the steel structure itself.	Figure 5 gives plots of water sorption data obtained from films of UW45 at 15°C.	Diffusion coefficients of about 5×10^{-14} m^2s^{-1} and water solubilities of about 0.5% were found from such experiments.	The experimental conditions were such that water was accessible to both surfaces of the films which were ~0.5 mm thick.

SURFACE PREPARATION

11.	In order to guard against the presence of a weak boundary layer and generally to optimise conditions for the adhesive bonding process, steel surfaces are first grit blasted underwater.	Once cleaned, the surface is immediately occupied

5

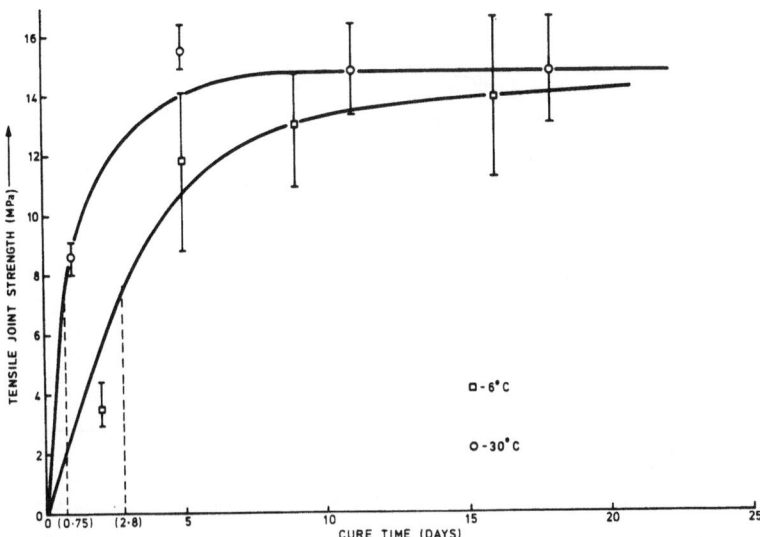

FIG. 3. HOW JOINT STRENGTH DEVELOPS WITH TIME AT DIFFERENT TEMPERATURES. (ADHESIVE UW45)

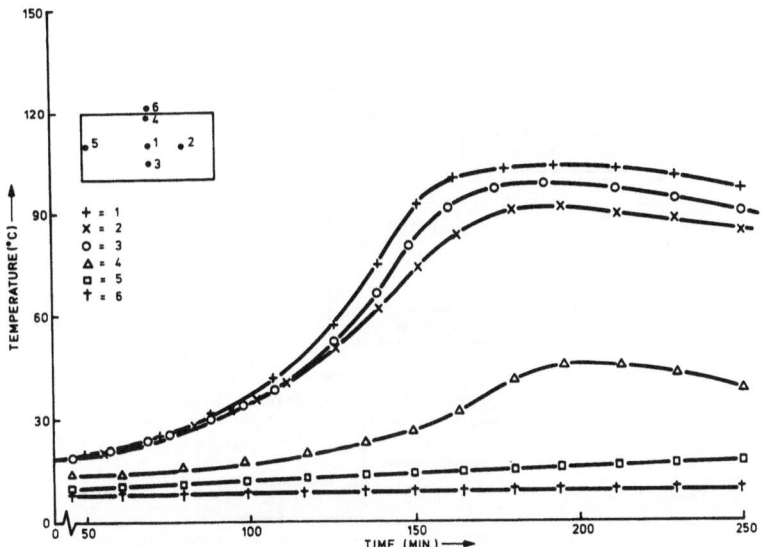

FIG. 4. EXOTHERM GENERATED WITHIN 1·4 x 10⁻² m³ UW45 WHEN OCCUPYING A MILD STEEL CONTAINER MEASURING 305 mm. x 305 mm. x 152 mm. AND SURROUNDED BY WATER AT 8° C.

6

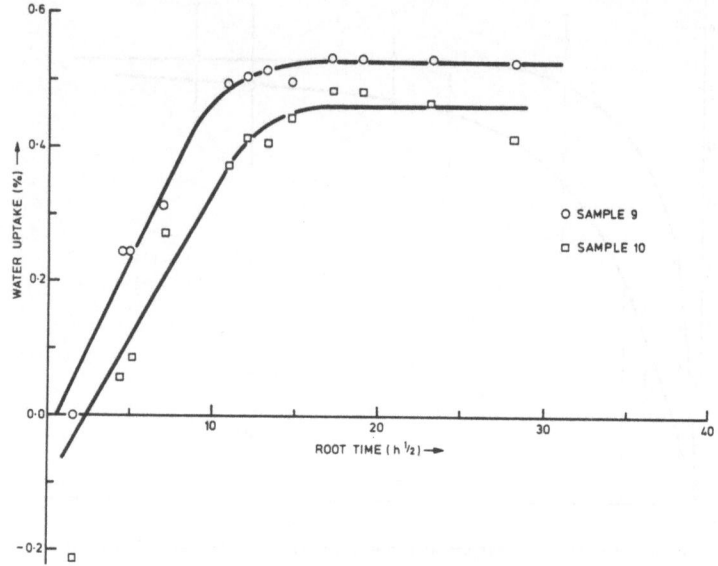

FIG. 5. WATER UPTAKE VERSUS ROOT TIME FOR UW45 AT 15°C.

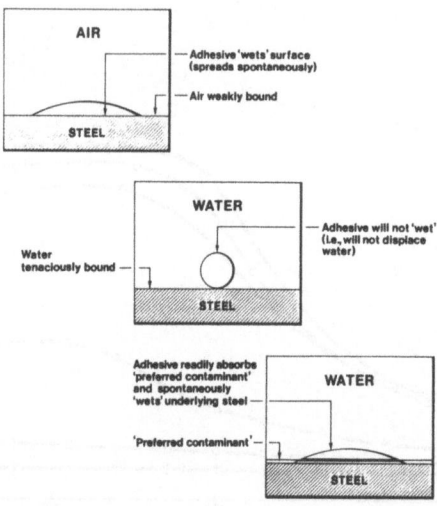

FIG. 6. ILLUSTRATING THE PROBLEMS OF "WETTING" STEEL UNDER
SEAWATER AND THE USE OF A SACRIFICIAL PRE-TREATMENT.

by water which is strongly adsorbed on the freshly exposed steel. Consequently, the adhesive, instead of spreading spontaneously as is normally the case on a clean substrate, fails to 'wet' the surface. The effective displacement of tenaciously adhering water is not easy to achieve; certainly it is very difficult to displace mechanically. However, the application of the so-called sacrificial pretreatment technology (SPT)[3] leads to the displacement of water from the cleaned steel and to the deposition of a hydrophobic film or preferred contaminant, over the surface. The nature of this deposited film is such that, although it is water repellant, it is also compatible with the adhesive to the extent that it may either be absorbed or displaced by it.

12. The net result of the application of this SPT is to re-establish underwater those favourable conditions commonly found in the atmosphere which allow spontaneous spreading of adhesive over the surface. Figure 6 illustrates this wetting problem and shows how the SPT is helpful in overcoming it.

ADHESIVE JOINT STRENGTH AND DURABILITY

13. The lower of the two histograms in Figure 7 indicates that where the favoured adhesive system is used immediately after grit blasting underwater to make steel/steel tensile butt joints the achievable joint strength is low ($\bar{\sigma}$ = 4.2 MPa) and, even more significantly, scatter is high (V = 32%). These results show the unpredictability of using purely mechanical methods for the displacement of water and the consequent incomplete wetting by adhesive of the substrate. With joints made in this way the failure mode appears to be exclusively interfacial between adhesive and steel adherend.

14. In the upper histogram the same figure also shows that where the displacement of water is energetically favourable after the use of the SPT then joint strengths are 3 to 4 times higher ($\bar{\sigma}$ = 16.5 MPa) and variation between results is greatly reduced to a value of around 5%. The failure of joints produced in this way is usually associated with cohesive failure within the adhesive layer. Such a low coefficient of variation compares well with those obtained when adhesive joints are made in the atmosphere.

15. In both histograms the 95% confidence limits are indicated by broken lines. It can be seen that six replicates were used to generate these data and for the most part all the results quoted in this paper are based on the debonding of six joints. For each joint the bonded area was 1290 mm^2 and glue line thickness was nominally 0.1 mm.

16. Although the bulk of the work was carried out with mild steel as the substrate, some other adherend materials were also investigated in a limited way, partly because of a general interest in them but partly also to establish the versatility of the sacrificial pretreatment approach to adhesive bonding.

8

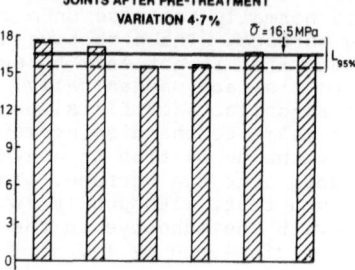

RESULTS FROM TYPICAL GROUP OF T.B.
JOINTS AFTER PRE-TREATMENT

VARIATION 4·7%

$\bar{\sigma}$ = 16·5 MPa

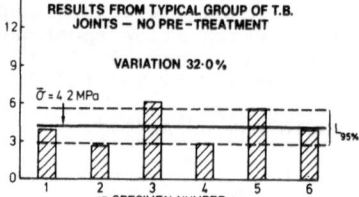

RESULTS FROM TYPICAL GROUP OF T.B.
JOINTS — NO PRE-TREATMENT

VARIATION 32·0%

$\bar{\sigma}$ = 4·2 MPa

FIG. 7. HISTOGRAM SHOWING THE EFFECT OF PRE-TREATMENT
ON JOINT STRENGTH.

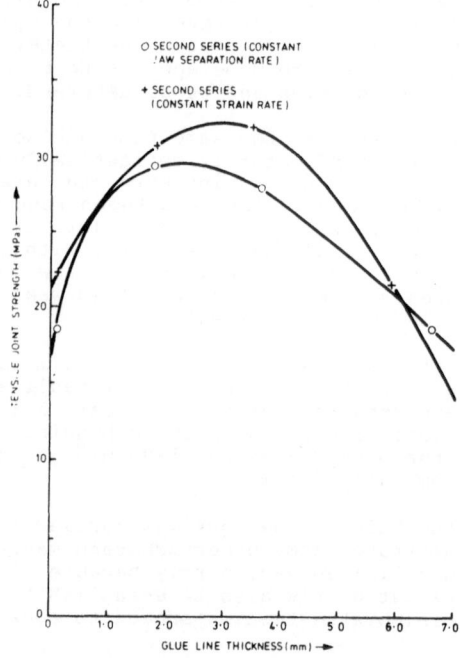

O SECOND SERIES (CONSTANT
JAW SEPARATION RATE)

+ SECOND SERIES
(CONSTANT STRAIN RATE)

FIG 8 RELATIONSHIP BETWEEN JOINT STRENGTH AND GLUE LINE THICKNESS

17. Aluminium bronze, aluminium alloy and a stainless steel
were included in this work. The results obtained are given in
Table 2 where it can be seen that for aluminium bronze and
aluminium alloy the presence of water during the joint making
process has no apparent effect on joint strength. However, in
the case of the stainless steel the presence of water resulted
in the loss of ~30% of the strength obtained when similar
joints were made in the dry. In addition, it is interesting to
note that in all cases the standard deviation is lower when
joints were made using the adhesive system underwater. A low
variation between results is obviously very important since
reliability and reproducibility are often crucial.

TABLE 2

JOINT STRENGTHS TO OTHER METAL SUBSTRATES

MATERIAL	TENSILE FAILURE STRESS, MPa	
	GB + SPT Bonded Underwater	GB ONLY Bonded in Air
Aluminium Bronze	10.5 ± 0.5	10.3 ± 1.0
Aluminium Alloy	8.0 ± 0.3	8.6 ± 0.5
Stainless Steel	19.1 ± 0.4	26.7 ± 1.2

18. Apart from this interest in other substrates there was
also an interest in a wider range of potential applications,
including the attachment of sacrificial anodes, and some
attention was also given therefore to the development of an
electrically conducting version of the adhesive.

19. Tensile butt joints were made up in the dry in the usual
way using metallic filled versions of the adhesive and these
test pieces were then used to assess the electrical properties
of the adhesives by measuring the voltage across each specimen
when a known and constant current was passed. Using the Ohm's
law relationship, the electrical resistance was measured and
this, together with a knowledge of glue line thickness and
bonded area, was used to calculate resistivity. A value of
20 ohm cm was found for the most promising version and since,
for cathodic protection devices, anode/cathode contact
resistance should be ⟨0.01 ohm and because the glue line
thickness in sub-sea applications may be typically around 5 mm,
a minimum, but readily achievable, bond area of about 0.1 m^2
would be required.

20. These joints were then broken in the Instron and the
results obtained are presented in Table 3 where they may be
compared with those made underwater in the usual way using the

preferred modified adhesive. Joint strength is clearly
unaffected by the presence of water during manufacture and it
is again shown that the coefficient of variation is
significantly lower when joints are made underwater. In both
cases the statistics are based on the best five of six
replicates. The rather higher failure stress found with these
joints may reflect a higher modulus associated with the use of
large quantities of metallic filler.

TABLE 3

UNDERWATER CONDUCTING ADHESIVE - JOINT STRENGTH TO MILD STEEL

	DRY	WET
$\bar{\sigma}$ (MPa)	23.1	22.2
S (MPa)	3.4	1.0
V (%)	14.6	4.4

21.   These results were useful in assessing the value of the
SPT approach but it was considered to be important that
additional experiments should be carried out which would
reflect more accurately conditions commonly found in the North
Sea.

22.   To this end, work was done in which the effect of glue
line thickness was investigated. Because of the
out-of-circularity associated with steel tubulars used in
offshore structures, annular gaps, for example, could not be
guaranteed to be less than about 6 mm. Figure 8 shows the
effect of increasing glue line thickness up to about this value
in standard steel/ steel butt joints against failure stress
under conditions of constant jaw separation rate, ie reducing
strain rate. Also presented in this Figure is a curve plotted
from results obtained with similar joints when jaw separation
rate was adjusted in order to maintain an approximately
constant strain rate in the adhesive layer. From the results
it was concluded that, over the range of glue line thicknesses
investigated, strain rate considerations were unimportant and
that joints with 6 mm glue lines were not much weaker than
standard joints with 0.1 mm glue lines. All these joints were
made in the dry.

23.   The effect of the lower temperatures found in the North
Sea on rate of cure has already been described (see Figure 3)
but in addition steel/steel lap and butt joints were made,
cured and debonded at +6°C and the results obtained are
compared in Table 4 with controls made, cured and debonded
under ambient temperature conditions (+21°C). It was expected
that lower temperatures, being equivalent to higher strain
rates, would be associated with higher apparent joint strengths
and the results obtained followed this pattern as can be seen
from Table 4.

TABLE 4

SHOWING THE EFFECT OF TEMPERATURE ON THE APPARENT

STRENGTH OF BONDED JOINTS

MAKING, CURING AND DEBONDING TEMPERATURE °C	LAP SHEAR FAILURE STRESS			TENSILE BUTT FAILURE STRESS		
	$\bar{\tau}$ MPa	S (MPa)	v (%)	$\bar{\sigma}$ MPa	S (MPa)	v (%)
+21	13.7	0.4	2.8	15.4	0.9	5.7
+6	14.3	0.5	3.6	18.3	1.3	7.3

24. The effect on apparent joint strength of curing under high hydrostatic pressures was also investigated. Although the maximum hydrostatic pressure likely to be experienced on the North Sea continental shelf is ~2 MPa, joints were made and then allowed to cure at a pressure of 7 MPa. The results were compared with those obtained from controls cured under ambient conditions (0.1 MPa) and the results appear in Table 5 where it can be seen that there were no important differences between the results obtained under the two experimental conditions.

TABLE 5

SHOWING HOW HYDROSTATIC PRESSURE APPLIED DURING CURE

INFLUENCES JOINT STRENGTH DEVELOPED BY UW45

HYDROSTATIC PRESSURE DURING CURE	TENSILE BUTT JOINT STRENGTH		
	$\bar{\sigma}$ (MPa)	S (MPa)	v (%)
7.0 MPa (absolute)	16.7	0.9	6.2
0.1 MPa (absolute)	15.5	1.6	10.3

25. Some of the more important work was directed at establishing the durability of joints made underwater and several experiments were carried out with this in mind.

26. Single overlap tensile shear joints were made using UW45 and these were exposed just below the waterline, suspended from a raft in Langstone Harbour. Results from these exposure trials, plotted as Figure 9, show that there was no loss of strength after 3 years. However, although both the adhesive/adherend interface and the adhesive itself remained

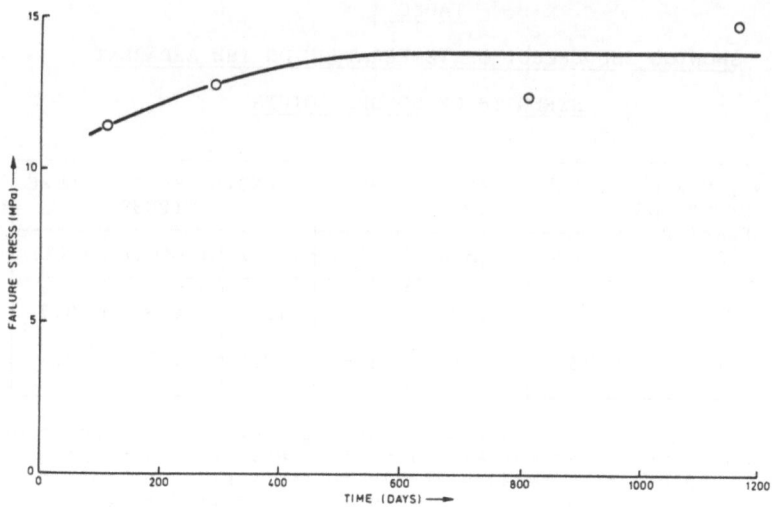

FIG 9 EFFECT OF LONG TERM EXPOSURE IN THE SEA ON TENSILE LAP SHEAR JOINTS.

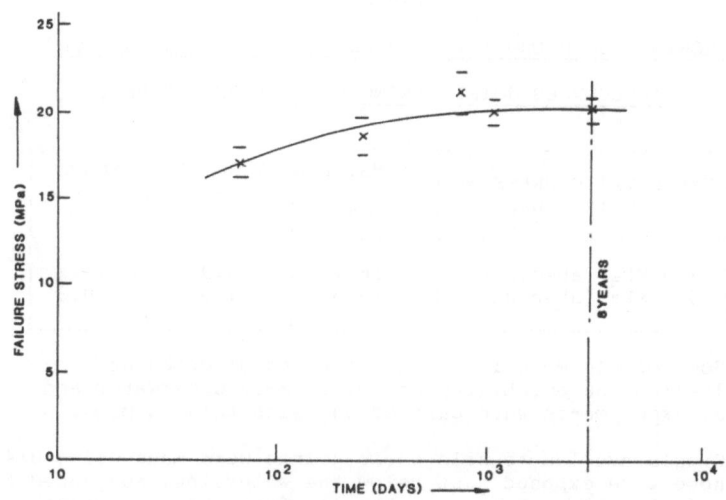

FIG. 10. EFFECT OF SEAWATER IMMERSION AT AMBIENT TEMPERATURE AND
PRESSURE ON TENSILE BUTT JOINTS.

unaffected by the environment, the rather thin section steel adherends were corroding badly and no further joints could be tested after the 1164 day withdrawals.

27. In addition tensile butt joints were exposed to seawater at room temperature in the laboratory and specimens have been withdrawn and debonded at intervals over a period approaching 8 years. The last specimens to be debonded showed no statistically significant loss of strength at the 95% confidence level. The results are given in Figure 10 where the 95% confidence limits are also indicated. The locus of failure in specimens at the 8 year withdrawal point remained 100% cohesive within the adhesive layer. Sufficient specimens remain for one further withdrawal.

28. Similar joints were exposed to seawater in a pressure pot under a hydrostatic pressure of 7 MPa, equivalent to a submerged depth of in excess of 640 m. Although this represented exposure to a hydrostatic pressure of more than 3 times that found on the continental shelf it can be seen from Table 6 that even after 6 years exposure there was no loss of strength although there appeared to be rather less cohesive failure within the adhesive layer and evidence of corrosion around the perimeter of the joints could be clearly seen.

TABLE 6

THE EFFECT OF SEAWATER AT HIGH PRESSURE (7 MPa) ON UW45

AND ON JOINTS MADE WITH IT

EXPOSURE PERIOD (days)	CAST ADHESIVE UTS (MPa)	TENSILE BUTT JOINT STRENGTH		TENSILE BUTT JOINT COHESIVE FAILURE (%)
		$\bar{\sigma}$ (MPa)	v (%)	
Initial Controls	22.8	16.7	10.1	95
14	25.5	13.6	18.0	83
45	18.9	15.8	13.8	100
121	21.3	16.7	7.6	87
2087	-	17.5	14.3	-
Final Controls (2087 days, 0.1 MPa)	-	16.9	14.5	70
Final Controls (2964 days, 0.1 MPa)	24.9	-	-	-

29. Dumbells of cast UW45 were also placed in the pressure pot at the same time and again there is no evidence of any loss of strength at the 121 day withdrawal period and controls, which were kept under seawater in the laboratory at an ambient pressure of ~0.1 MPa, showed no loss of strength after nearly 8 years immersion.

30. In addition, to investigate further the long term durability of adhesive joints made underwater with the UW45 adhesive, stress rupture tests were also carried out. To establish the effects of continuous loading, tensile butt joints were made and fitted in creep machines as shown in Figures 11 and 12. Chains of 3 joints were assembled in a jig designed so that, in the event of a failure in one joint, the load would be taken up by an otherwise loose stainless steel linkage system and the load would be maintained on the remaining specimens.

31. Chains of specimens were exposed under seawater at 20°C at 40, 30, 25, 20 and 10% of their short term failure stress. In Figure 13 the loading level expressed as a percentage of short term joint strength is plotted against time to failure. Although failure at the 40% level occurred within a few hours, specimens subjected to the 20% loading level survived about 3 years before they failed. This performance is believed to be good even when compared with adhesive joints stressed in the atmosphere. In general, it is not recommended that structural adhesive joints should be subjected to continuous loading in excess of about 20% of their short term strength. Those exposed to the 10% loading level were expected to survive for an extremely long time and so these were withdrawn after about 3 years and debonded in the Instron when it was found that they failed at a mean stress level of 18.2 MPa, a value some 7% higher than the original control value.

CONCLUSIONS

32. An adhesive and a substrate cleaning and surface preparation method for mild steel have been developed to a stage where usefully strong and durable adhesive joints may be reproducibly made under seawater. The surface preparation method involves the application of novel technology in which a specially developed sacrificial pretreatment is deposited onto the steel surface before the adhesive itself is applied. The adhesive may certainly be applied at temperatures down to ~6°C and the adhesive will cure to give strong joints at temperatures down to +3°C.

33. The durability of steel/steel joints made with the UW45 adhesive is such that unstressed specimens have survived for almost 8 years under seawater at the relatively high temperature of ~20°C without any noticeable loss of strength. Stress rupture tests indicate that even when loaded in tension, rather than in the preferred shear mode, joints survived for 3 years when loaded underwater to 20% of their short term strength at 20°C.

15

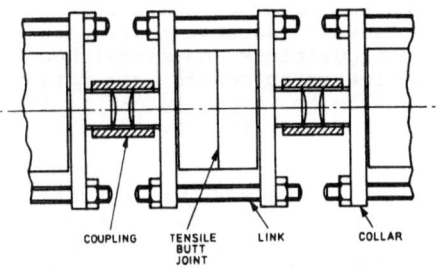

FIG. 11. CONSTRUCTION OF STRESS RUPTURE TEST PIECE CHAIN.

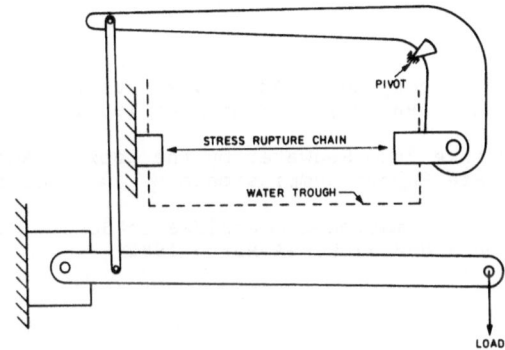

FIG. 12. CREEP MACHINE ARRANGEMENT.

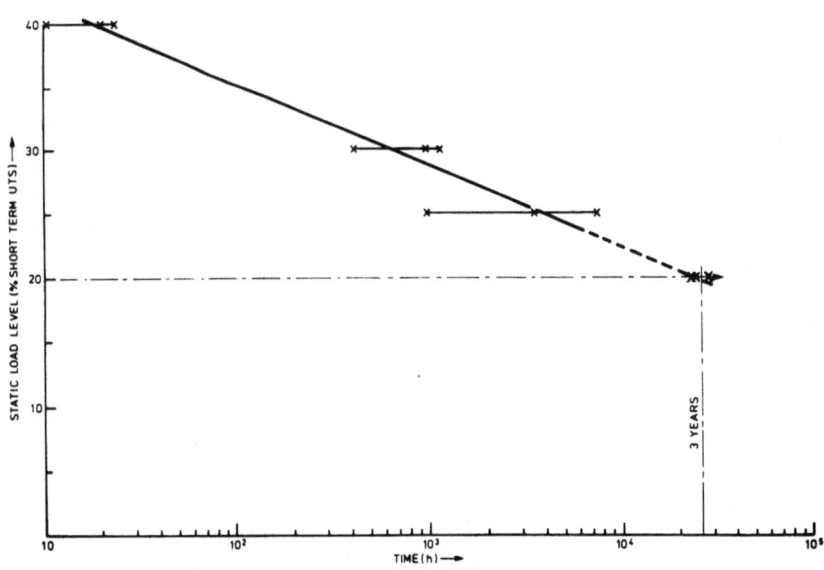

FIG 13. STRESS RUPTURE RESULTS FOR TENSILE BUTT JOINTS UNDER SEAWATER.

34. Water diffusion and water uptake studies on the bulk adhesive give results which suggest that, when used in realistic repair configurations and sandwiched between steel with a low joint perimeter to bonded area ratio, joint strength degradation attributable to water ingress is likely to be negligible over the expected lifetime of North Sea production rigs.

35. Although the bulk of the work was concerned with mild steel as the substrate material there are indications that the technology may be used successfully with other metallic substrates such as aluminium bronze, aluminium alloy and stainless steels. An electrically conducting version was also developed which may find application in the attachment of sacrificial anodes.

REFERENCES

1.   BOWDITCH M R. Epoxy Resin Adhesives. UK Patent Specification 1 598 432, 23 September 1981.

2.   COMYN J. Uptake from Seawater by the Epoxide Adhesive UW45. (Private Report commissioned by ARE, Holton Heath.)

3.   BOWDITCH M R. Formation of Metal/Resin Bonds. UK Patent Specification 2 083 377B, 24 March 1982

Chapter 2

SOME ASPECTS OF SILANE TECHNOLOGY FOR SURFACE
COATINGS AND ADHESIVES

C Kerr and P Walker

SCT Materials, AWRE, Aldermaston, Berkshire

1.   INTRODUCTION

Organofunctional silanes of the general formula $R-Si(OR')_3$ are
used extensively as adhesion promotors in the glass-fibre reinforced
plastic and composite industries, and the literature reflects this
interest [1-5]. This early and continued use is in recognition of the
fact that the physical properties of glass reinforced composites are
severely reduced in the presence of water. The deleterious effect of
water on the adhesion of surface coatings to the substrate and the bond
strength of adhesives, particularly to metals and glass, is now fully
recognised [6-8] and the composite technology and experience is now being
applied in these fields.

Silane adhesion promotors, or coupling agents represent a
relatively simple and low technology approach to the problem of loss of
adhesion of organic materials in the presence of water, either under
immersion conditions, or high humidity. The purpose of this present paper
is to review the chemistry of the silanes relevant to their use as
adhesion promotors, briefly review the mechanism of adhesion promotion
by reference to published theories and present data on the use of silanes
in and with surface coatings and adhesives on metallic and other substrates. The major emphasis will be on the benefits arising from the use
of silanes in mitigating the effect of water on bond strength.

2.   SILANE COUPLING AGENTS

Silane coupling agents are monomeric species of the general
formula $R-Si(OR')_3$, where R represents an organofunctional group and
(OR') is a hydrolysable ester group. Silicon chemistry allows the
preparation of a large number of compounds of this type to be synthesised
and the commercially available range is extensive. The silanes used in
the work reported in this paper are listed in Table 1.

2.1   Reactions of Interest in Coupling

The reactions of interest may be summarised as:

1.   Hydolysis of ester group

$$R-SiX_3 + 3H_2O \xrightarrow[\text{catalyst}]{\text{pH}} R-Si(OH)_3 + 3HX \qquad \text{eq. (1)}$$

2.   Hydrogen Bonding at Surface

$$R-Si(OH)_3 + HO-Si \rightarrow \qquad \text{eq. (2)}$$

TABLE 1

Silanes Described in the Paper

Y - methacryloxy propyltrimethoxysilane $\quad\quad\quad$ . . . . MAMS

$$CH_2 = \underset{\underset{CH_3}{|}}{C} - \underset{\underset{O}{||}}{CO} (CH_2)_3 \ Si(OCH_3)_3$$

B - (3, 4, epoxycyclohexyl) ethyltrimethoxysilane $\quad$ . . . . ECMS

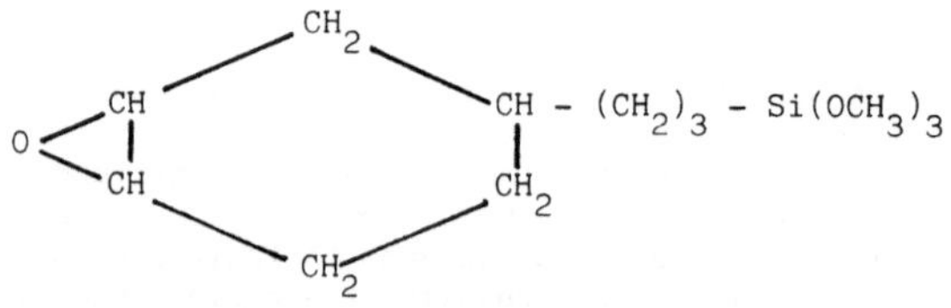

Y - glycidoxy propyltrimethoxysilane $\quad\quad$ . . . . GPMS

$$CH_2 - CH - CH_2 - O(CH_2)_3 \ Si(OCH_3)_3$$

Y - mercaptopropyl trimethoxysilane $\quad\quad$ . . . . MPS

$$HS(CH_2)_3 \ Si(OCH_3)_3$$

Y - aminopropyltriethoxysilane $\quad\quad$ . . . . APES

$$NH_2 \ CH_2 \ CH_2 \ CH_2 \ Si(OC_2H_5)_3$$

N - beta (aminoethyl) - gamma aminopropyltrimethoxysilane

. . . . AAMS

$$NH_2 \ (CH_2)_2 \ NH(CH_2)_3 \ Si(OCH_3)_3$$

3.   Condensation with the surface

$$R-Si(OH)_3 + HO - \overset{\displaystyle(\ \ (}{\underset{\displaystyle(\ \ (}{Si}}) \rightarrow R-Si-O-Si + H_2O \qquad\qquad \text{eq. (3)}$$

4.   Polymerisation

$$R-Si(OH)_3 + R-Si(OH)_3 \xrightarrow[\text{Low pH}]{\text{High or}} \begin{matrix} OH & OH \\ | & | \\ R-Si-O-Si-R+H_2O \\ | & | \\ OH & OH \end{matrix} \qquad \text{eq. (4)}$$

5.   Reaction with the polymer

$$- C \overset{\displaystyle\diagdown\diagup}{\underset{O}{-}} C + R - NH_2 \longrightarrow HO - \overset{|}{C} - \overset{|}{C} - NHR$$

primary amino
group on silane   eq. (5)

　　　　Equation 5 is typical of the many possible reactions between
silanes and organic polymers, the nature of which will be determine by
the functional groups present in both polymer and the silane. A more
detailed treatment of probable reactions with epoxide and urethane groups
is given by Walker [9].

2.2   Mechanisms of Adhesion Promotion

　　　　The mechanisms by which silanes bond to the surface of metals
and glass substrates has attracted a great deal of attention and, reflect-
ing the early technological interest in glass reinforced plastics, the
major research effort has been concentrated on the mechanism of bonding
to glass.

　　　　Detailed reviews of the topic have been published by
Erickson [10] and Rosen [11].

　　　　The oldest and most widely known theory is the "Chemical Bond
Theory" in which it is postulated that trialkoxysilanes chemically bond
to the substrate through interaction of the alkoxy group with surface
silanols on the glass and to the organic present via reaction of the
organofunctional group with reactive species in the polymer. Strong interfacial bonds of the order of 50-100 kcal/mol are formed. The difficulties
with this theory are that some silanes appear to function as adhesion
promotors although chemically non-reactive with the polymer and it does
not necessarily explain the improved wet adhesion as metallosiloxane bonds
to metal surfaces are susceptible to hydrolysis.

　　　　The "Deformable Layer Theory" postulates a plastic interface,
providing a zone in which stresses between the high modulus surface and
the relatively low modulus organic layer may be relieved without bond,

rupture, thus internal stresses are minimised. It has been suggested that the interfacial zone is too thin to allow stress relaxation in the required time scale.

The "Surface Wettability Theory", applied initially to filled systems argues that complete wetting of the filler particles by the resin would improve the adhesion by physical adsorption. Although the increased bond strength would exceed the cohesive strength of the polymer it is difficult to see how in the competition with water and other weakly bonded layers, physical adsorption is likely to provide much reinforcement to a bond where chemical bonding is also present.

In the "Restrained Layer Theory" it is postulated that a region of intermediate modulus is formed to transfer stress from the high modulus surface to the low modulus polymer. Silanes function by tightening up the polymer structure at the interface and simultaneously provide silanol groups for bonding to the substrate. The need for stress relaxation due to thermal shrinkage between polymer and filler or substrate is not accounted for. In effect this theory postulates a similar reaction between polymer and substrate as the "Chemical Bond Theory" while advancing a different mechanism for its observed action.

The "Reversible Hydrolytic Bond Mechanism Theory" is largely a combination theory combining features of the "Chemical Bond Theory" with the rigid interface of the "Restrained Layer"and allowing for the stress relaxation properties of the "Deformable Layer". It proposes the reversible breaking and re-forming of stressed bonds between coupling agent and substrate which allows stress relaxation without loss of adhesion in the presence of water since SiOH has a strong hydrogen bonding capacity, water is eliminated from the surface and a high initial bond strength is achieved. It has also been argued that when Si–O metal bonds are broken by the intrusion of water, the bond is capable of reforming with consequent recovery in adhesion.

2.3   Requirements For Adhesion Promotion

In order to function as an adhesion promotor a silane must fulfil certain requirements, these may be summarised as:

1)   The organofunctional groups on the silane must be capable of reaction with functional groups in the polymer.

2)   Chemically reactive groups must be available in the polymer (coating or adhesive).

3)   There must be some mechanism for the silane to concentrate at, or migrate to, the substrate/polymer interface.

4)   The silane (used as an additive) should not be depleted by water, solvent or pigment interactions.

5)   The reactivity of the silane organofunctional group should be such that reaction with the polymer is possible before the polymer's reactive groups are consumed or immobilised in the curing reaction.

In view of the comment made earlier it is possible that (2) is not mandatory.

2.4   Methods of Use

Silanes may be used in three main ways:

2.4.1   As Pretreatment Primers

Here the silane is applied from solvent solution to the substrate and functions as a primer in its own right.

2.4.2   As a Formulated Primer

Here the silane is incorporated into a solvent solution of epoxide or melamine resin. Examples of formulated primers are given by Plueddeman [12].

2.4.3   As an Additive

In this case the silane is added directly to a single pack system at the manufacturing stage or to the appropriate component of a two pack system.

The latter method, in which the silane is added to the polymer system during manufacture is almost universely desirable but may introduce storage difficulties. Depletion of silane may occur by reaction with the polymer, absorption by the pigment, reaction with water or with oxygenated solvents.

3.   Silanes and Surface Coatings

Although it may be considered desirable to improve the initial adhesion of a coating to the substrate, the real problem lies in increasing the adhesion under water soaked conditions and obtaining high recovered adhesion. The data on surface coatings is largely covered under the three headings initial, wet, and recovered defined as:

> Initial - measured after cure under normal environmental conditions.

> Wet   - measured under the environmental conditions of interest and before recovery eg immediately after water immersion.

> Recovered - measured after a period of recovery under normal environmental conditions.

Two methods of measurement have been used; Torque Shear [13] in which cylindrical test specimens are bonded to the coating surface and twisted off using a recording torque spanner and support table coupled via a universal joint, and the Direct Pull-off method [14] in which coated discs are bonded between cylindrical specimens and the resultant assembly broken on an Instron Universal Test Machine in direct tension.

3.1   Silanes as Pretreatment Primers

The results of experiments in which the silane AAMS was applied from a 1$^{w/}$o solution in various solvents are shown in Table 2. It can be seen that for a two pack polyurethane paint applied to a degreased aluminium substrate considerable improvements in both initial and wet adhesion were obtained, the presence of water either in the solvent solution or as a wash gave the highest values. Although the actual values recorded by the two test methods were different, the general agreement on effect, was good.

TABLE 2

Effect of Solvent in Pretreatment Primer

(Polyurethane Paint on Aluminium, Silane AAMS)

Solvent	Torque Shear		Direct Pull-off	
	Initial MPa/area detach	Wet MPa/area detach	Initial MPa/area detach	Wet MPa/area detach
No Silane (degreased)	32.3/100	10.0/100	12.6/100	4.1/100
Dry methylethyl-ketone	36.6/0–100	15.2/100	23.0/0–80	8.4/30–100
Dry petroleum ether	40.4/0–30	25.3/100	25.9/100	11.2/50–90
Dry methyl ethyl ketone/Water Wash	47.5/0	41.4/0+50	22.0/40	25.5/80
Acetone/Water Solution	39.3/0–60	44.5/20	21.1/60–100	19.2/90

The initial, wet and recovered adhesion values for the polyurethane paint applied over different silanes on an aluminium substrate and subjected to a 1500 hour water immersion test are shown in Table 3. The improvements in initial bond strength on both the degreased and grit-blasted substrates were marked, with the greatest improvement on the degreased substrate. A considerable improvement in wet adhesion was also recorded and the recovered bond strengths were also considerably higher.

TABLE 3

Effect of Water Immersion on Polyurethane Paint on Aluminium

(Direct Pull-off) (1500 hours)

Silane Treatment/Preparation	Initial MPa/area detached	Wet MPa/area detached	Recovered MPa/area detached
Degreased only	12.6/100	3.8/100	9.9/100
MAMS	32.3/30	10.1/30	21.8/30
AAMS	26.3/40	11.1/30	22.8/30
Gritblasted only	28.6/10	8.5/100	13.6/80
MAMS	33.7/0	13.0/20	22.4/40
AAMS	34.0/0	14.9/20	22.0/30

Data on the same paint applied to silane treated mild steel is shown in Table 4 from which it can be seen that although the initial bond strength, particularly on the AAMS treated surface was considerably higher than that of the non-silane controls, the wet adhesion on the degreased panels was little improved. On the gritblasted surface a considerable increase in bond strength over that of the control was observed. The recovered bond strengths were considerably higher.

TABLE 4

Effect of Water Immersion on Polyurethane Paint on Steel

(Direct Pull-off) (1500 hours)

Silane Treatment/Preparation	Initial MPa/area detached	Wet MPa/area detached	Recovered MPa/area detached
Degreased only	16.7/100	5.7/100	6.8/100
MPS	25.2/30	5.4/100	12.1/100
AAMS	38.2/0	7.4/90	12.9/90
Gritblasted only	25.7/40	11.8/95	20.8/60
MPS	32.2/5	23.7/10	25.3/0
AAMS	36.7/0	22.8/30	29.2/0

Reference to Table 5 will show the data recorded on the polyurethane paint applied to mild steel after exposure to cyclic humidity under condensation conditions. In every case there was a considerable improvement in initial, wet and recovered bond strength, particularly on the AAMS treated panels.

TABLE 5

Effect of Cyclic Humidity, 42-48°C on Polyurethane Paint on Steel

(Direct Pull-off) (500 hours)

Silone Treatment/Preparation	Initial MPa/area detached	Wet MPa/area detached	Recovered MPa/area detached
Degreased only	15.4/100	3.5/100	6.6/100
Gritblasted only	35.9/10-40	15.2/100	18.8/100
MAMS degreased	34.6/0-90	17.8/10-50	24.0/20
ECMS degreased	29.2/20-90	25.5/0-10	27.6/0-2
GPMS degreased	31.7/50-70	26.9/0-10	28.6/0
MPS degreased	34.0/0-90	21.9/0-2	30.1/0
AAMS degreased	40.0/0	21.6/80	31.0/0

Data on the two pack epoxide paint applied to silane treated aluminium is shown in Table 6. Both silanes investigated produced improvements in initial, wet and recovered bond strength on both types of surface preparation

On the mild steel substrate (Table 7) all the silanes produced an improvement in the initial bond strength on both the degreased and gritblasted surfaces. The improvement was greatest on the AAMS treated panels. The wet bond strength values were also considerably greater, with the exception of GPMS treated gritblasted substrate. In every case the recovered bond strength was appreciably higher. Data on the epoxide paint after exposure to cyclic humidity is shown in Table 8. In every case the silane treatment resulted in an improvement in the initial, wet and recovered adhesion, particularly the latter.

3.2   Silanes as Additives

Additions of silanes in quantities up to 0.4 $^w/o$ on total resin content were made just prior to application of the paint. Initial work in which additions up to 5$^w/o$ showed that additions in excess of 1$^w/o$ either showed no greater effect than the lower addition or resulted in a reduced bond strength.

TABLE 6

Effect of Water Immersion on Epoxide Paint on Aluminium

(Direct Pull-off) (1500 hours)

Silane Treatment/Preparation	Initial MPa/area detached	Wet MPa/area detached	Recovered MPa/area detached
Degreased only	21.4/90	5.7/100	11.2/100
MAMS	30.2/0	12.0/30	19.7/50
AAMS	31.2/0	11.5/30	20.5/40
Gritblasted only	28.5/30	8.5/100	13.7/80
MAMS	31.8/10	13.3/40	21.6/50
AAMS	32.5/0	13.0/40	25.0/30

TABLE 7

Effect of Water Immersion on Epoxide Paint on Mild Steel

(Direct Pull-off) (1500 hours)

Silane Treatment/Preparation	Initial MPa/area detached	Wet MPa/area detached	Recovered MPa/area detached
Degreased only	19.9/60	7.2/100	11.0/100
ECMS	27.4/20	17.3/100	21.8/90
GPMS	30.1/0	14.3/100	15.6/100
MPS	27.2/40	11.0/30	16.4/100
AAMS	32.0/0	28.1/100	29.2/10
Gritblasted only	25.9/40	9.2/100	21.0/100
ECMS	27.7/0	16.3/30	31.7/60
GPMS	31.9/0	8.3/70	27.0/100
MPS	31.2/0	17.2/10	30.1/45
AAMS	33.6/0	25.3/50	27.9/40

TABLE 8

Effect of Cyclic Humidity, 42-48°C, Epoxide Paint on Mild Steel

(Direct Pull-off) (500 hours)

Silane Treatment/Preparation	Initial MPa/area detached	Wet MPa/area detached	Recovered MPa/area detached
Degreased only	19.5/40	17.4/100	18.1/100
Gritblasted only	25.9/30	21.8/100	23.1/80
MAMS Gritblasted	31.2/0	26.4/60	29.2/20
ECMS Gritblasted	30.4/0	28.3/0-2	30.0/20
GPMS Gritblasted	31.9/0	26.5/0-20	31.1/0
MPS Gritblasted	31.2/0	24.8/10-30	33.3/0
AAMS Gritblasted	33.6/0	26.4/50-80	30.6/0

Reference to Table 9 will show the data recorded for the polyurethane paint applied to aluminium. Clearly the silanes used as additives produced an increase in initial, wet and recovered bond strength. There appeared to be an advantage in using the AAMS at the higher level of addition. Similar data (Table 10) is shown for the epoxide paint on aluminium. In this case the improvements in wet and recovered bond strength were particularly marked.

4.   SILANES AND ADHESIVES

Two methods of test have been used to evaluate the effect of silanes on adhesives, both in direct tension ie butt tensile. Where the technical metals and uranium were used the specimens consisted of bolts of head area 0.2 in^2 bonded together in pairs to form a doublet. In the case of beryllium and frangible glass the variant technique known as the "Sandwich" test was used, in this case discs of the chosen substrate were bonded between the heads of the bolts, and, the resultant specimen cured in an alignment pressure jig. In both cases the specimens were broken on an Instron Universal Testing Machine at a cross-head speed of 1mm/min.

4.1   Silanes as Pretreatments

The silanes were applied as a 2% solution in an alcohol/water solution. The data for a structural two pack polyurethane adhesive on aluminium, stainless steel and mild steel is shown in Table 11. It can be seen that not all the silanes were effective on all three metals.

TABLE 9

Effect of Water Immersion on Polyurethane Paint on aluminium

(Torque Shear) (1500 hours)

Silane Addition/Treatment	Initial MPa/area detached	Wet MPa/area detached	Recovered MPa/area detached
Degreased only	29.1/90	9.3/100	6.2/100
0.4% MPS	33.8/90	12.6/100	31.7/100
0.1% AAMS	36.3/100	10.1/100	14.6/100
0.2% AAMS	37.7/30	20.0/100	20.0/100
Gritblasted only	33.1/100	31.2/5-10	36.9/20-60
0.1% AAMS Gritblasted	44.3/0	39.7/0	48.3/0
0.2% AAMS Gritblasted	45.7/0	39.2/0	46.9/0

TABLE 10

Effect of Water Immersion on Epoxide Paint on Aluminium

(Torque Shear) (1500 hours)

Silane Addition/Treatment	Initial MPa/area detached	Wet MPa/area detached	Recovered MPa/area detached
Degreased only	27.6/100	6.7/100	16.6/100
0.1% MPS	32.1/80	32.6/0-20	44.1/20-80
0.2% MPS	45.9/0-5	33.1/0-20	43.1/0-60
0.1% AAMS	45.8/0	41.4/0	46.9/0
0.2%	47.2/0	41.4/0	42.8/15
Gritblasted only	32.7/40	24.0/20-100	29.2/20-60
0.2% MPS Gritblasted	45.1/0	38.6/0	49.7/0
0.2% AAMS Gritblasted	44.4/0	41.4/0	48.3/0

The silane, AAMS, was probably the most effective all-round bond strength promotor, particularly on aluminium and stainless steel. MPS performed well on all but the aluminium substrate and it is possible that the result on this substrate indicates a less than optimum application.

Similar data for a polyamide cured structural epoxide adhesive is shown in Table 12. AAMS performed well on all substrates, as did MPS. The silanes were least effective on mild steel.

TABLE 11

Effect of Various Silanes on Bond Strength of
Polyurethane Adhesive to Metals
(degreased substrate) (as pretreatment)

	NONE	GPMS	MPS	APES	AAMS
Aluminium					
Bond Strength	19.0	15.7	9.8	16.9	24.8
Coefficient of Variation	19.1	7.5	27.2	15.6	17.4
Site of Failure	AS/CF	AS/CF	AS/CF	AS/CF	CF/AS
Stainless Steel					
Bond Straight	37.8	32.9	42.5	23.7	40.0
Coefficient of Variation	12.6	33.0	21.5	37.7	25.6
Site of Failure	AS/CF	AS	CF/AS	AS	CF/AS
Mild Steel					
Bond Strength	17.5	14.1	24.8	11.4	19.7
Coefficient of Variation	31.1	13.2	24.8	15.0	12.8
Site of Failure	CF/AS	AS	CF	AS	AS/CF

AS   adhesion failure from substrate
CF   cohesive failure in the adhesive

TABLE 12

Effect of Various Silanes on Bond Strength of
Epoxide Adhesive to Metals
(degreased substrate) (as pretreatment)

METAL	NONE	GPMS	MPS	APES	AAMS
Aluminium					
Bond Strength . . . . MPa	29.0	35.6	35.4	32.9	35.0
Coefficient of Variation . %	20.3	19.8	18.4	22.0	18.1
Site of Failure	AF/CF	CF	CF	CF	CF/AS
Stainless Steel					
Bond Strength . . . . MPa	22.1	18.2	36.6	–	26.9
Coefficient of Variation . %	15.8	18.6	18.5	–	15.9
Site of Failure	AF/CF	AS/CF	CF/AS	–	CF/AS
Mild Steel					
Bond Strength . . . . MPa	13.6	16.9	15.8	13.6	18.1
Coefficient of Variation ½ %	25.5	41.3	10.1	26.5	32.3
Site of Failure	AF	AF	AF	AF	AF

4.2   Comparison of Silanes as Pretreatments and Additives

Reference to Table 13 will show the results of an initial
experiment in which the silane, MPS, was used either as a 1-8 $^{w}/o$ solution
in alcohol/water or as an additive at the same range of additions, based
on the total resin solids. In every case the measured bond strength on
the pretreated specimens was greater than on those bonded with the
additive adhesives. The former indicated that the primer concentration
was not important, the latter, that the bond strength decreased with
increasing silane level. In the experiments reported in Tables 14, 15 and
16 the silanes were used at the 2 $^{w}/o$ level in the primer and 1 $^{w}/o$ as
an additive.

Reference to Table 14 which shows the data for the polyurethane
adhesive applied to stainless steel prepared by an initial gritblasting.
Only GPMS produced any marked improvement in bond strength when used as a
pretreatment. In all cases the use of the silane as an additive resulted
in a lower bond strength than the control. It should be noted that all
the specimens pretreated with silane failed cohesively, the additive
specimens showed some component of adhesion failure.

Table 15 shows that the epoxide adhesive failed cohesively in
all cases. On mild steel (Table 16) all the silanes used as pretreatments
produced an increase in bond strength, they were less effective as
additives.

TABLE 13

Effect of MPS on Bond Strength of
Epoxide Adhesive to Aluminium
(Butt Tensile)

	% MPS							
	1		2		4		8	
	Primer	Additive	Primer	Additive	Primer	Additive	Primer	Additive
Bond Strength MPa	47.7	43.7	48.7	35.1	44.4	31.0	46.1	29.1
Coefficient of Variation	4.0	4.5	15.0	17.0	7.0	11.6	23.8	7.9
Site of Failure	MCF	MCF	MCF	MCF	MCF	MCF	MCF	MCF

MCF - mainly cohesive failure in adhesive

TABLE 14

Effect of Various Silanes on Bond Strength of
Polyurethane Adhesive to Stainless Steel
(gritblasted substrate)

	Silane								
	None	GPMS		MPS		AAMS		APES	
		Primer	Additive	Primer	Additive	Primer	Additive	Primer	Additive
Bond Strength MPa	46.3	52.3	35.4	43.7	36.5	47.8	43.8	39.8	42.9
Coefficient of Variation	8.9	5.9	11.4	7.7	13.6	4.7	6.5	5.3	13.3
Site of Failure	CF	CF	AS	CF	AF	CF	AS/CF	CF	AS/CF

TABLE 15

Effect of Various Silanes on Bond Strength of
Epoxide Adhesive to Stainless Steel
(gritblasted substrate)

	Silane							
		GPMS		MPS		AAMS		APES
	None	Primer	Addi-tive	Primer	Addi-tive	Primer	Addi-tive	Primer	Addi-tive
Bond Strength MPa	63.0	65.0	63.0	66.0	54.6	66.3	61.6	59.4	59.9
Coefficient of Variation	10.8	5.0	12.5	4.8	11.9	6.4	7.5	10.7	10.4
Site of Failure	CF	CF	CF	CF	CF	CF	CF	CF	CF

TABLE 16

Effect of Various Silanes on Bond Strength of
Epoxide Adhesive to Mild Steel
(gritblasted substrate)

	Silane							
		MAMS		GPMS		MPS		AAMS
	None	Primer	Addi-tive	Primer	Addi-tive	Primer	Addi-tive	Primer	Addi-tive
Bond Strength MPa	38.3	48.8	21.8	52.0	39.5	48.2	42.0	52.3	41.8
Coefficient of Variation	7.4	6.3	19.9	4.5	11.8	7.6	13.4	2.5	7.9
Site of Failure	MCF	MCF	AS	CF	AS	CF	AS	CF	CS

5.   MITIGATION OF WATER EFFECTS

        Specimens of the structural polyurethane and epoxide adhesives
on stainless steel were exposed to 100% humidity at room temperature for
periods up to 24 months. Specimens were removed from the high humidity
at various time intervals and the joints broken before any recovery could
take place.

        The data for the polyurethane adhesive is shown in graphical
form in Figure 1. It can be seen that both GPMS and MPS produced a
marked increase in bond strength retention, the specimens pretreated with
GPMS retained 49% of their original bond strength and MPS 35% compared
with 15% for the non-silane control.

The epoxide adhesive data is shown in Figure 2. Both silanes produced an improvement in bond strength retention. The GPMS specimens retained 35% of their original bond strength and the MPS, 30%. The non-silane controls failed completely before testing, having lost 89% of their original bond strength in the first 12 months.

6.   SILANES AND URANIUM

Uranium is a particularly reactive metal the water vapour corrosion of which is inhibited by oxygen [15]. The result is that a polymeric coating which allows water ingress but prevents or retards oxygen diffusion to the surface is likely to cause rapid and total surface corrosion, with complete loss of bond strength of the superimposed polymer layer whether coating or adhesive.

The results of two experiments in which depleted uranium substrates in bolt form were bonded with a structural urethane adhesive (Mat.S370) and a flexible urethane (Mat.S373) and exposed at 60°C and 10% RH are shown in Figures 3 and 4 respectively. Figure 4 shows that the use of AAMS produced a marked improvement in the initial bond of the structural adhesive and more importantly, improved the bond strength retention, effectively more than doubling the life of the bond. The results for the flexible adhesive using the same silane are less promising but nevertheless represent an improvement in the working life of the bond.

7.   SILANES AND BERYLLIUM

The results of a single experiment in which the silane MPS was applied as a pretreatment primer to beryllium and tested by the sandwich technique is shown in Table 17. The use of MPS resulted in a marked increase in initial bond strength, and more importantly, an increase in bond strength retention after exposure to high humidity. The retention values for the silane were 80% compared with 65% for the non-silane. The "wet" bond strength of the latter was only 65% of that of the former.

8.   SILANES AND FRANGIBLE GLASS

Frangible glass is an ion-exchanged glass in which the outer layer is in compression and the inner in tension, penetration of the outer layer results in franging of the glass to form small regular parti-cles, not unlike, but smaller, than a shattered windscreen. The material finds use as a device for the rapid remote release of the contents contained within it. Initial experiments with a wide range of adhesives indicated that although high initial bond strengths could be achieved using epoxide adhesives, rapid loss of bond strength occurred on exposure to water vapour. The results of a preliminary survey of the effect of silanes on the initial bond strength are shown in Table 18. Clearly not all the silanes were effective as adhesion promotors. A short term exposure to high humidity showed that although the use of APES and AAMS as pretreatment primers gave the highest initial strength, the bond deterio-rated rapidly on exposure to high humidity. The results of a longer term exposure of the amine cured epoxide adhesive, 815/TETA/DT075, at 100% RH and 30°C are shown in Figure 5. The retained bond strength values after exposure for 90 days were 30% for MAMS 57% for ECMS, 68% for GPMS and 58% for the MPS. The non silane control retention value was 24%. Data on a longer term test at 100% and RT on MPS

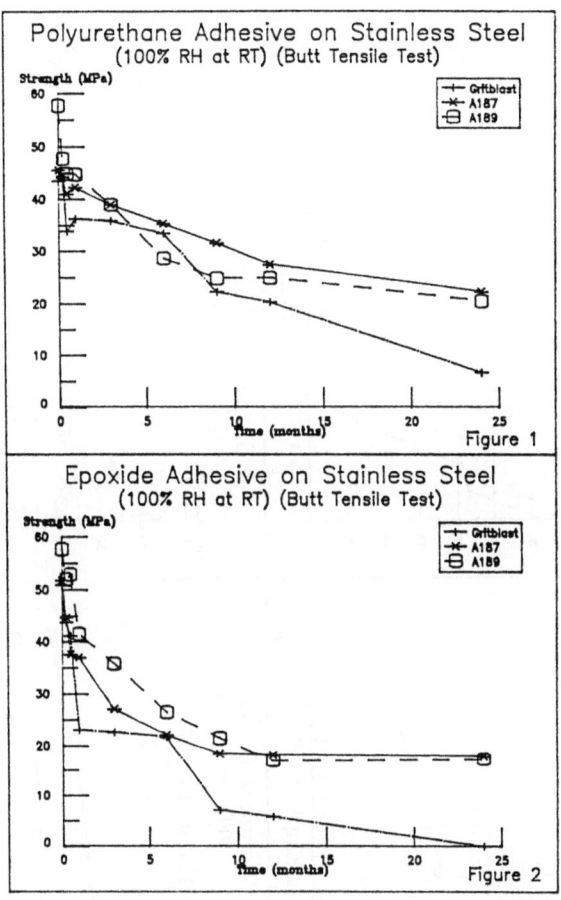

Figure 1

Figure 2

treated frangible glass is shown in Figure 6. The non-silane control failed completely between 100 and 200 hours. The MPS treated specimens retained 42% of their original bond strength after 90 days exposure.

34

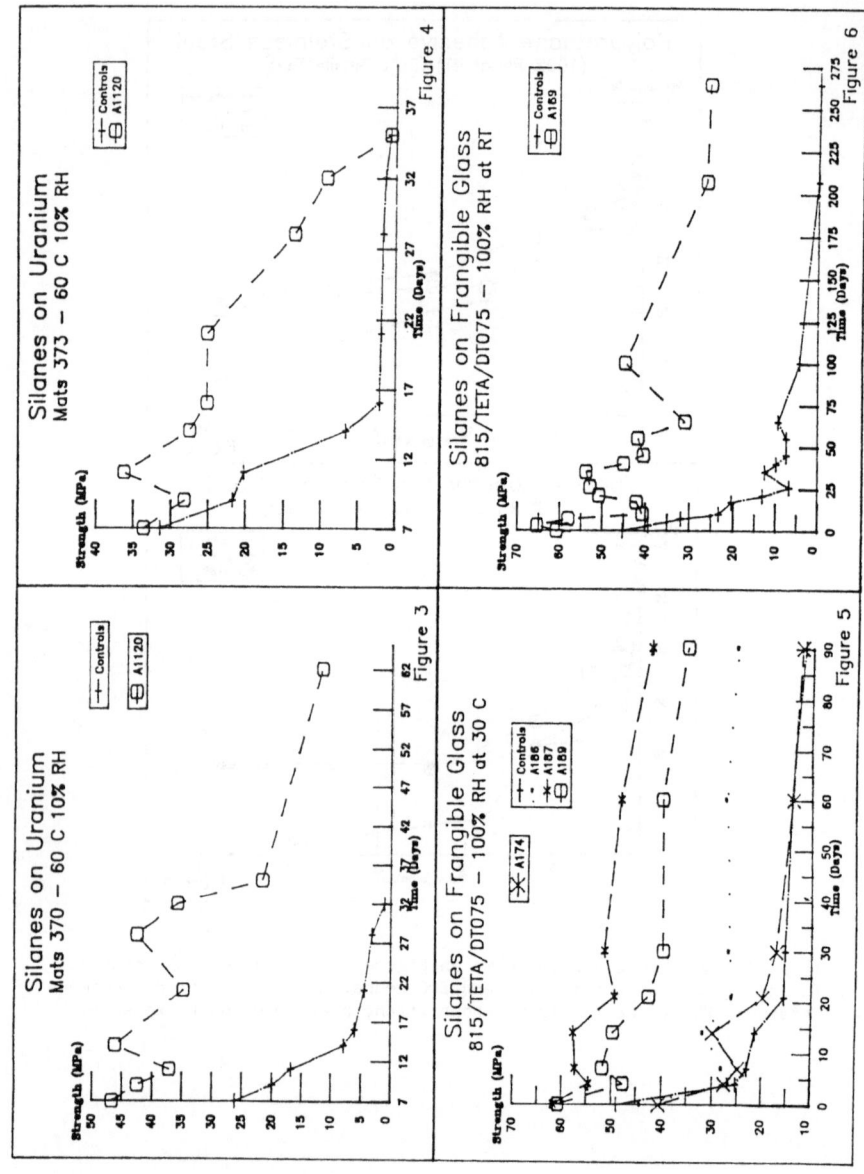

Silanes on Uranium
Mats 373 — 60 C 10% RH

Figure 4

Silanes on Uranium
Mats 370 — 60 C 10% RH

Figure 3

Silanes on Frangible Glass
815/TETA/DTO75 — 100% RH at RT

Figure 6

Silanes on Frangible Glass
815/TETA/DTO75 — 100% RH at 30 C

Figure 5

TABLE 17

Effect of Silane on Bond Strength of
Epoxide Adhesive to Beryllium
(degreased) (sandwich specimen) (MPS)

	No Silane Control After 30 days at 100% RH		Silane Treatment Control After 30 days at 100% RH	
Bond strength . . . . . . .MPa	39.9	25.9	49.9	39.5
Coefficient of Variation .%	6.7	44.4	16.7	8.6
Site of Failure	AS	AS	CF	CF/AS

TABLE 18

Effect of Silanes on the Bond Strength of
Epoxide Adhesives to Frangible Glass
(Butt tensile-sandwich technique) (14 days cure at RT)

	Silane						
	None MPa/ Fail- ure	MAMS MPa/ Fail- ure	ECMS MPa/ Fail- ure	GPMS MPa/ Fail- ure	MPS MPa/ Fail- ure	APES MPa/ Fail- ure	AAMS MPa/ Fail- ure
Amine Cured Epoxide	29.3	13.5	19.8	30.5	36.8	45.0	43.4
(815/TETA/DT075)	AFG	AFG	AFG	AFB	AFB	AFB	AFB
Polyamide Cured Epoxide	26.9	7.9	13.9	30.1	25.8	26.5	29.6
(825/125)	AFG	AFG	AFG	AFB	AFB	AFB	AFB

AFG - adhesion failure from the glass.

AFB - adhesion failure from the bolt.

9.   DISCUSSION

The work on surface coatings indicates that when silanes are
used as pretreatment primers the solvent or solvent blend from which they
are deposited can have a major effect on both the initial and wet bond
strength. The data in Table 2 in which AAMS was deposited from a range
of solvents clearly indicates that the presence of water either within
the solvent mixture or as a water wash after deposition is important if
the maximum benefit is to be achieved. It may well be that the water
wash after deposition of the film effectively removes the less polymerised
outer layer of the deposited film leaving only a thinner more highly
polymerised layer in contact with the substrate. Clearly, there are
aspects other than the choice of silane which are important, parameters
which may also be important are the thickness of the deposited layer and
probably also the pH of the silane solution. Credence to the latter is
provided by work reported by Boerio and Williams [16] in which the silane
was deposited from solutions of pH 8.0 to 12.5,a considerable difference
in retained bond strength under water soaked conditions was noted.

In general the results obtained from surface coatings of the
polyurethane and epoxide types show that considerable improvements in
initial, wet, and recovered adhesion can be achieved. The improvements
in wet and recovered adhesion are particularly marked, and the silanes
perform well on all three substrates tested. The highest bond strength
values were obtained from the use of MPS and AAMS.

Major improvements were obtained when the silanes were used as
either pretreatment primers or additives, and the data indicates that
when used as additives at the $0.2^{w}/o$ level slightly higher bond strengths
were achieved.   This is in direct contradiction to the results obtained
from the adhesive specimens where the use of a silane as an additive
showed less improvement in bond strength than the same silane used as a
pretreatment. It would seem possible that this difference arose from
either the reduced mobility of the silane in the 100% solids high viscosity
adhesives as opposed to the high solvent content, low viscosity coatings,
or depletion of the silane by reaction with the adhesive polymers by
acting as a curing agent. In view of the relatively short time interval
between addition of the silane and application of the adhesive, the latter
mechanism appears to be improbable. Work reported on surface coatings
[17] indicated that silanes used as additives in polyurethane and epoxide
paints still provided an increase in bond strength after two years storage.

The use of silanes as pretreatment primers for adhesives showed
that considerable improvements in retained bond could be obtained on a
variety of substrates and paralleled the work on surface coatings.

10.   CONCLUSIONS

From the work reported it is concluded that the use of
organosilanes as pretreatments for and as additives to, surface coatings
of the two pack polyurethane and epoxide type can result in a marked
improvement in the intial, wet and recovered bond strengths to aluminium
and mild steel substrates. Improvements in the initial bond strength of
adhesives to metallic substrates may also result from the use of silanes
as pretreatment primers and a considerable improvement in bond strength
retention can be achieved on metal and glass substrates. Although effective as additives in adhesives this approach is less effective than the
use of silanes as pretreatments.

11.   <u>REFERENCES</u>

[1]   Bjorksten, J and Yalger, L.L., Mod. Plastics, 29, 124, 1952.

[2]   Sterman, S., and Marsden, J.G., Ibid, 43, 1966.

[3]   Sterman, S.S., et al., Paper 16, RPG Conference on New and
       Improved Resin Systems, London, September 1973.

[4]   Plueddeman, E.P., and Stark, G.L., "Surface Modification of
       Fillers and Reinforcements in Plastics", Paper at 32nd Annual
       SPI. RPC. Conference. Section 4C, p1 1977.

[5]   Rosen, M.R., Journal of Coatings Technology, 50, 644, 70, 1978.

[6]   Falconer, D., MacDonald, N., and Walker, P.,Chem. Ind. July
       1964.

[7]   Walker, P., Official Digest, 37, 1561, 1965.

[8]   Funke, W., JOCCA; 68, 9, 229, 1985.

[9]   Walker, P., Journal of Coatings Technolgoy, 52, 670, 49, 1980.

[10]   Erickson, P.W., and Plueddeman E.P., "Composite Materials"
       Vol 6, Academic Press, New York, 1974.

[11]   Rosen, M.R., Journal of Coatings Technology, 56, 644, 70, 1978.

[12]   Plueddeman, E.P., Progress in Organic Coatings, 11, 297, 1983.

[13]   Holloway, M.W., and Walker, P., JOCCA, 47, 5, 812, 1964.

[14]   Paper submitted by the Joint Services R&D Committee on Adhesion
       and Accelerated Weathering, JOCCA, 46, 4, 1963.

[15]   Orman, S., and Walker, P., JOCCA, 48, 233, 1965.

[16]   Boerio, F.J., and Williams, J.W., Applications of Surface
       Science, 7, 19, 1981, North Holland Publishing Co.

[17]   Walker, P., JOCCA, 65, 436, 1982.

Chapter 3

ESTER POLYMERS AND THEIR INTERACTION WITH ALUMINA
STUDIED BY INELASTIC ELECTRON TUNNELLING SPECTROSCOPY

J. COMYN, C.C. HORLEY, R.G. PRITCHARD and R.R. MALLIK

Faculty of Science, Leicester Polytechnic, Leicester LE1 9BH

1.   INTRODUCTION

Inelastic electron tunnelling spectroscopy (IETS) allows one to record the vibrational spectrum of a monolayer of organic molecules adsorbed upon the metal oxide of a metal/metal oxide/metal thin film sandwich - an IET junction (1-3). In this work the junctions were aluminium, aluminium oxide, adsorbed organic layer, and lead. If a d.c. bais voltage is applied to the junction, a net current flows due to electrons tunnelling from one metal to the other. Approximately 1% of these electrons tunnel inelastically and so lose energy to molecular oscillators in the oxide or adsorbed layer. This produces a small increase in conductance of the IET junction at a bias voltage V given by the relationship $V = h\nu/e$ where h is Planck's constant, e is the electronic charge and ν the frequency of the oscillator. This increase is more easily seen as a peak in a plot of the second derivative of the voltage with respect to current (d^2V/dI^2) against bias voltage V. IETS is usually performed with the junction at a temperature 4.2 K, mainly to reduce thermal broadening of the IET lines. Both Raman and infra-red (IR) vibrational modes may be activated.

IETS is particularly appropriate for the study of organic compounds used as adhesives or adhesion promotors adsorbed on aluminium oxide as aluminium is a widely studied and commercially important material for adhesive bonding. Other workers have used the technique to investigate both epoxides and their mixtures with aliphatic amines (4), cyanoacrylates (5) and silane coupling agents (6) adsorbed on aluminium oxide. However, none of these incorporated polymers in IET junctions, although in some cases small molecules were used which then polymerized on the oxide surface.

We have been able to incorporate some polymers containing ester groups into IET junctions both by solution doping (7) and by plasma polymerization (8) and it is the results of these investigations which are described here. Polyvinylacetate (PVA) and polymethylmethacrylate (PMMA) were the polymers used in solution doping, whilst these and polyacrylic acid (PAA), polyethylacrylate (PEA) and polyacrylonitrile (PAN) have been incorporated by plasma polymerization.

Plasma polymer films are deposited by introducing a controlled flow of monomer vapour into a sealed, evacuated chamber within which a radio frequency (RF) glow-discharge may be established. Energy imparted to the monomer molecules generates active centres for polymerization. A polymer film may then be formed on any substrate placed within the discharge. The structure and composition of such films depend mainly on the operating conditions of the plasma, the geometry of the chamber and RF electrodes, and to a lesser degree the nature and flow-rate of the monomers (9, 10, 11). Films deposited for the applications mentioned above, and for IR spectroscopy, are generally of the order of several μm thick, highly crosslinked, and relatively pinhole-free (12).

The present work will demonstrate that plasma polymer films with a thickness of the order of 0.1 nm may be incorporated as the insulating layer of Al/Al-oxide/insulator/Pb, inelastic electron tunnelling spectroscopy junctions. Hence plasma polymerization offers a new and complementary doping technique for polymers in addition to the liquid-phase method. An obvious advantage of plasma polymerization is that it precludes the need for solvents and polymerization may be performed in inert atmospheres.

Polymers with ester side groups find uses in a number of adhesives including PVA adhesives for paper cardboard and wood, and reaction setting modified acrylic adhesives for metal structures.

2. EXPERIMENTAL

Polymers for solution doping were prepared by free radical addition polymerization (7). Vinyl acetate was polymerized in solution in acetone at 65°C using azobisisobutyronitrile (AZBN) as the initiator and had a viscosity average molar mass of $90000 \overset{+}{-} 5000$. PMMA was prepared by suspension polymerization in water at 80-85°C using AZBN as initiator; viscosity average molar mass was $27000 \overset{+}{-} 2500$.

Details of our IET spectrometer and of our general procedures for preparing IET junctions by liquid or vapour phase doping have been published previously (13), and a brief summary is given here.

Firstly, the aluminium base electrode typically $\sim$ 200 nm thick is evaporated onto a clean glass substrate (a microscope slide) through a brass mask, which allows the geometry of the electrode to be defined. This procedure is carried out within a vacuum evaporation chamber at a pressure of $\sim 10^{-5}$ Torr. On removal from the vacuum chamber a thermal oxide approximately 1-2 nm thick is rap idly formed on the aluminium. The slide is then placed on a rotary spinner where a solution of the desired polymer is poured onto the slide and the excess spun off.

The slide is replaced in the vacuum plant and the Pb counter electrode is evaporated typically to a thickness of $\sim$ 300 nm, thus completing the IET junction. The junction is then immersed in liquid helium, and vibrational modes in the range 0 to 500 mV (0 to 4025 cm^{-1}) are investigated with our spectrometer.

When using plasma polymerization as a means of doping junctions, immediately after the evaporation of the aluminium base electrodes the glass slide was removed from the vacuum chamber and positioned in the plasma apparatus. Here a 13.56 MHz RF power supply was capacitatively coupled to a parallel plate, flat-bed glow-discharge configuration within the chamber of a modified Nanotech autocoater 300 vacuum plant.

Prior to polymerization the chamber and its contents were cleaned with an argon plasma for several minuters, and then gettered by the evaporation of a thin layer of lead, approximately 100 nm. This procedure reduces the amount of water vapour and atmospheric contamination adsorbed on the internal surfaces of the chambers to an acceptable level. The glass slide was then mounted on a holder mid-way between a pair of circular discharge plates, and held at a floating potential by means of two PTFE insulating poles. These poles were shielded by stainless steel cylinders to prevent any energy being dissipated to them.

It was convenient to supply the RF oscillator with a d.c. voltage of 200-300 V at a d.c. current of 100 mA; no excessive heating of the electrodes occuurred at an RF output corresponding to these values. The RF power supplied to the electrode system may be estimated roughly by simultaneous measurement of the RF current and potential difference associated with the latter. The figure obtained of $\sim$ 10-20 W was used throughout this work.

Polymer film thickness was essentially controlled by the monomer flow-rate. However, since discharge times were in general quite short (of the order of a few seconds) the amounts of monomer used were small, so prohibiting any accurate determination of the flow-rates. For this reason it is more convenient to quote the dynamic chamber pressures during discharge and the exposure times rather than absolute flow-rates. Over a more prolonged period it has been ascertained that the corresponding flow-rates lie in the approximate range 0.1 to 0.3 cm^3 min^{-1} for the range of chamber pressures used in the present work. Typical values of chamber pressures and exposure times were 50-100 m Torr, and 5-30 s respectively but actual values for the individual monomers used are given in Table 1.

TABLE 1   -   Operating conditions employed for the plasma polymerization of the monomers used for IET junctions.

Monomer	Chamber Pressure/m Torr	Approx. Exposure Time/s	Approximate IET Junction Resistance/Ω
Vinyl acetate	100	5	70
Ethyl acrylate	100	10	80
Acrylic acid	50	30	100
Methylmethacrylate	100	15	200
Acrylonitrile	50	30	50

All monomers were distilled from over calcium hydride and then vacuum degassed by three freeze-thaw cycles in liquid nitrogen. Once the monomer flow had been instigated and a stable pressure attained, the RF discharge was established thus initiating polymerization.

Thicker plasma polymer films were prepared on KBr discs for IR spectrophotometry. This was done by reducing the plate separation to 30 mm, increasing the RF power level by adjusting the d.c. input parameters to 200 mA at 700 V, and increasing the monomer exposure times and flow rates (typical values were several tens of minutes, and between 0.5 and 0.7 cm^3).

3.   RESULTS AND DISCUSSION

Figure 1 shows a set of blank IET spectra obtained from five IET
junctions on the same glass slide, the oxidised Al base elctrodes having
been subjected to an RF glow-discharge in laboratory air.   The spectra
are similar to those normally obtained from IET junctions with undoped,
thermally grown oxides.   Low junction resistances, approximately 30Ω, and
very weak spectral features due to CH, OH and $\overline{COO}$ vibrational mdes (see
Table 2) suggest that traces of atmospheric hydrocarbons, water, and
possibly formic acid are adsorbed on the oxide surface.   Levels of
contamination are commensurate with those normally observed in undoped IET
junctions.

3.1   Solution Polymerized Polymers

Figure 2 shows the IET spectrum for PVA,doped from a 0.025% w/v
solution of acetone and peak assignments are given in Table 3.   The
spectrum has two strong peaks at 1444 and 1596 cm^{-1}.   These are widely
believed to be due to the symmetric, and asymmetric stretching vibrations
of the carboxylate resonance hybrid $\overline{COO}$.

A probable explanation for the presence of these peaks is acid- or
base-catalysed hydrolysis of ester groups by hydrated alumina to give
acetic acid and polyvinylalcohol.   This is possible since aluminium oxide
is amphoteric, and hence both basic and acidic sites may hydrolyse the
ester.   It is known that one of the products, acetic acid, is strongly
adsorbed on alumina as the acetate ion, which in turn gives rise to the
characteristic carboxylate peaks (14, 15).

In addition to the presence of acetate-like features, very strong
aliphatic CH stretching modes are observed $\sim$ 2912 cm^{-1}.   In a semi-
empirical study of monobasic aliphatic acids adsorbed upon alumina,
Walmsley and Nelson have shown that some interdependence exists between
increasing chain length, and the intensity of the CH stretching vibrations
(16).   A similar effect has also been observed for long chain surfactant
molecules (17).   We offer no firm explanation for this observation, but
note that in general the IET spectra obtained for polymeric samples in
this work exhibit similarly strong CH modes.   This may again be due to
the orientation of the C-H bonds relative to the oxide surface.

The IET spectrum of adsorbed PVA also shows evidence of low energy
polymeric skeletal vibrations of the carbon backbone.   The above
polymeric features are presumably due to adsorbed polyvinylalcohol.

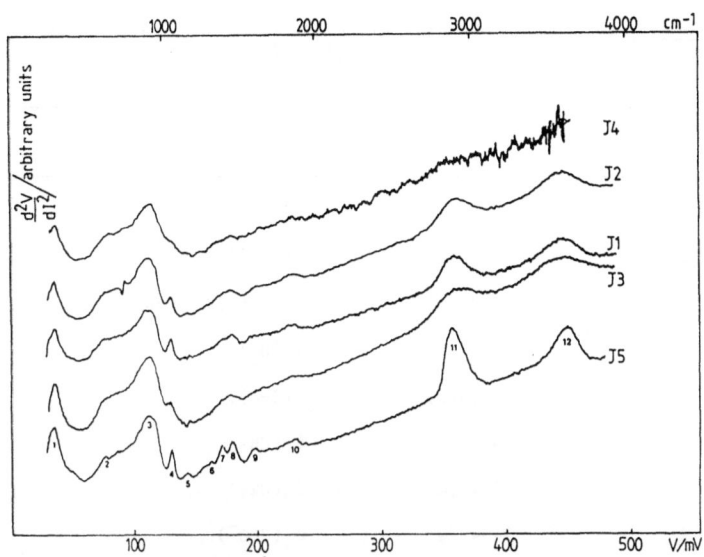

FIGURE 1 — IET spectra of undoped devices after exposure to an RF discharge in laboratory air. (Approx. pressure 50–100 m Torr)

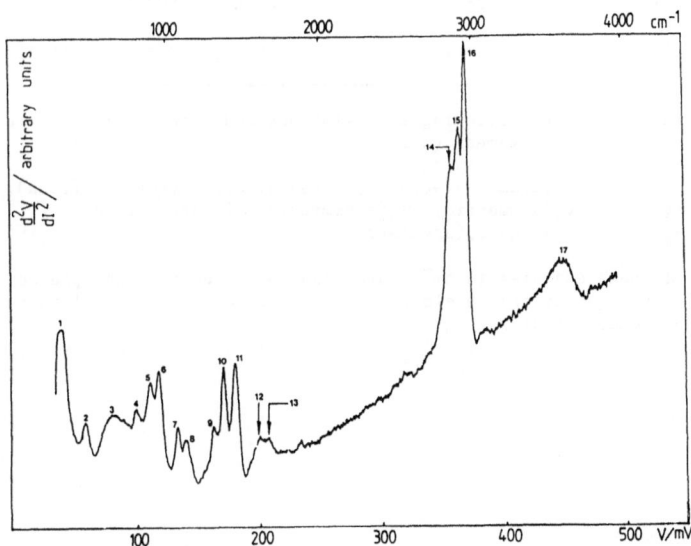

FIGURE 2 — IET spectrum of PVA doped from a ∿ 0.025% w/v solution in acetone

TABLE 2   -   Assignment for blank IET spectra obtained after devices exposed to glow-discharge in air.

Peak No.	cm^{-1}	Assignment
1	282 (m)	Al phonon
2	613 (w)	$r(COO^-)$ and OH bend
3	919 (m)	Al-O bulk phonon
4	1049 (wm)	$r(CH_3)$
5	1145 (w)	(C-O)?
6	1307 (vw)	$\delta_s(CH_3)$
7	1371 (w)	$\delta_a(CH_3)$
8	1444 (w)	$\nu_s(COO^-)$
9	1597 (vw)	$\nu_a(COO^-)$
10	1855 (vw)	Al-O (2x) overtone
11	2879 (m)	$\nu(CH_2)$, (CH_3)
12	3621 (m)	$\nu(OH)$, surface hydroxyls

Footnote: The following abbreviations and symbols are used in assignment tables.

w(weak), m(medium), s(strong), sh(shoulder), v(very), r(rock), ν(stretch)ν_s(symmetric), ν_a(asymmetric), δ(deformation), δ_s(symmetric) and δ_a(asymmetric).

All peak energies in cm^{-1} have been corrected for the superconducting energy gap of the Pb electrode by substraction of 8 cm^{-1} in the assignment tables.

TABLE 3 - Assignments for IET Spectrum of PVA

Peak No.	cm^{-1}	Assignment
1	290 (s)	Al phonon
2	444 (w)	r(COO$^-$)
3	621 (w)	OH bend
4	782 (wm)	skeletal (C-C)?
5	879 (ms)	ν(C-C)
6	936 (ms)	Al-O bulk phonon
7	1057 (m)	r(CH$_3$)
8	1113 (m)	skeletal ν(C-C-O) or ν(C-O)
9	1299 (w)	δ_s(CH$_3$)
10	1363 (ms)	δ_a(CH$_3$)
11	1444 (ms)	ν_s(COO$^-$)
12	1597 (w)	ν_a(COO)$^-$
13	1661 (w)	ν(C=O)
14	2863 (vs)	ν(CH$_3$), ν(CH$_2$) modes
15	2912 (vs)	
16	2952 (vs)	
17	3605 (m)	ν(OH)

Figure 3 shows the IET spectrum of PMMA doped from a 0.05% w/v solution PMMA in acetone, and the relevant peak assignments are given in Table 4. The most significant features are again the strong symmetric and asymmetric stretching vibrations of the carboxylate anion at 1460 cm^{-1} and 1621 cm^{-1} respectively, and a reduction in intensity of the carbonyl peak at 1677 cm^{-1} relative to that of the IR spectrum. This would again suggest acid or base-catalysed hydrolysis of the ester groups to the carboxylate. As proprosed by Hall and Hansma (14), it is believed that carboxylic acids dissociate on the Al-oxide surface. They are subsequently adsorbed by the formation of a bidentate symmetrical bridging complex between carboxylate anions and the Al ions in the oxide. As a result, strong carboxylate peaks are observed in IET spectra. Evidence of these peaks in Figure 3 would indicate that carboxylate side groups of the hydrolysed PMMA may also be adsorbed in this manner.

Detection of the methanol by-product would be difficult since it is known that lower aliphatic alcohols are suitable solvents for IETS doping in that they are not strongly adsorbed on aluminium oxide. Presumably, the alcohol would either evaporate from the oxide, or be pumped away in the vacuum system prior to the evaporation of the Pb electrode.

3.2 Plasma Polymers

The IET spectra of the plasma polymers all show good evidence for polymerization including a reduction in intensity the vinyl absorption usually observed at 1640 cm^{-1} and the presence of polymeric features such as low energy skeletal modes associated with vibrations of the carbon backbone. Strong aliphatic CH stretching, rocking, and deformation modes typical of polymeric samples in IETS are also present.

The IET spectrum of plasma polymerized VA (not shown) is very similar to that of the solution polymerized substance. Similar adsorbed species are therefore present in both cases; this would suggest that no significant differences exist between the composition of the adsorbed plasma film and that of liquid-phase doped PVA. In both cases there occur peaks at about 1435 and 1590 cm^{-1} assigned to symmetric and asymmetric vibration of the carboxylate group. Significantly these peaks are absent from the transmission IR spectrum (not shown) of a thicker sample of plasma polymerized vinyl acetate deposited on a KBr disc, so showing that the formation of these groups depends on the presence of the

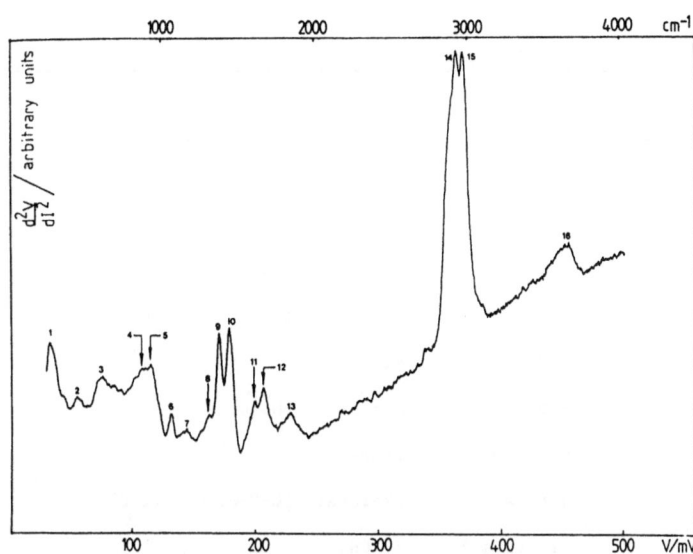

FIGURE 3 — IET spectrum of PMMA doped from a ∿ 0.05% w/v solution in acetone

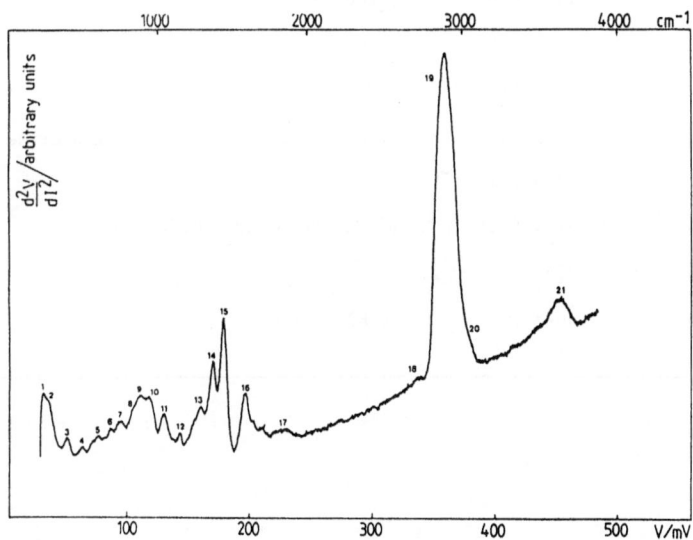

FIGURE 4 — IET spectrum of plasma polymerized ethylacrylate

TABLE 4 - Assignments for IET spectrum of PMMA

Peak No.	$^{-1}$	Assignment
1	274 (s)	Al electrode phonon
	355 (sh)	
2	452	r(COO$^-$) in plane
3	613 (ms)	r(COO$^-$) out of plane and OH bend
4	879 (wm)	ν(C-C)
5	936 (m)	Al-O bulk phonon
6	1065 (wm)	r(CH$_3$)
7	1178 (w)	skeletal (C-C-O) or (C-O)
8	1323 (w.sh)	δ_s(CH$_3$)
9	1387 (s)	δ_a(CH$_3$)
10	1460 (s)	ν_s(COO$^-$)
11	1621 (m)	ν_a(COO$^-$)
12	1678 (m)	ν(C=O)
13	1855 (w)	Al-O (2x) overtone and/or Al hydride stretch
	2750 (vw)	
14	2944 (vs)	ν(CH$_2$) and ν(CH$_3$) modes
15	2968 (vs)	
16	3678 (wm)	ν(OH) surface

alumina surface. The very weak peak at 3073 cm^{-1} is assigned to $= CH$, this may be due to the residual monomer or due to unsaturation arising from termination by disproportionation.

The same group is possibly present as a very weak shoulder (No.20 in Fig.4 and Table 5) in plasma polymerized ethylacrylate. Cleavage of this polymer on alumina would now yield ethanol and acrylic acid units. The former is commonly used as a solvent in the liquid-phase doping of IET junctions; it seems to readily escape from junctions without leaving any discernable trace. Acrylic acid units would account for the detection of the carboxylate peaks (Nos.3, 5, 6, 15 and 16). The spectra of PEA and PAA (Fig .5 and Table 6) are very similar both in assignments and particularly in the relative intensities of the peaks assigned to ν_s (COO$^-$) and ν_a (COO$^-$) vibrations. This implies that the degree of ester group cleavage of PVA adsorbed on alumina is high.

The spectrum of plasma PMMA (not shown) is very similar to that for the solution polymer, which implies that the plasma and bulk polymerized species have similar structures when adsorbed on alumina. Again with the plasma polymer, peaks appear which are assigned to the carboxylate group. The intensities of the symmetric and asymmetric carboxylate peaks are comparable with those for PEA, so implying that the level of ester group cleavage is high here. The IR spectra of plasma and conventionally polymerized PMMA (not shown) are closely similar.

Figure 6 and Table 7 give details of the IET spectra of plasma polyacrylonitrile. Peak 13 can be assigned to the nitrile group, but there are also peaks due to the symmetric and asymmetric modes of the carboxylate group. These probably arise from partial hydrolysis of nitrile groups on the hydrated alumina surface by the following reactions.

$$- CN \xrightarrow{H_2O} - CONH_2 \xrightarrow{H_2O} - COONH_4 \longrightarrow - COO^-$$

The low intensities of the carboxylate peaks in comparison with spectra for other polymers indicate that hydrolysis is incomplete, however there are no features in the IET spectrum which can be assigned to amide groups or the ammonium ion. Hence it seems that a fraction of the nitrile groups react, and react to the end of the sequence shown above.

We have therefore demonstrated that polymers containing ester groups undergo chemical reactions at an aluminium oxide surface (which was grown in air and is therefore hydrated) to produce carboxylate ions.

TABLE 5 - Assignments for plasma polymerized ethyl acrylate

Peak No.	cm^{-1}	Assignment
1	258 (m)	$\delta(C-C-O)$?
2	290 (m.sh)	Al phonon
3	411 (w)	$r(COO^-)$ in plane?
4	516 (w)	
5	613 (w)	$r(COO^-)$ out of plane
6	702 (w)	$\delta_s(COO^-)$
7	766 (w)	
8	839 (w.sh)	
9	887 (m)	$\nu(C-C)$
10	952 (m)	Al-O bulk phonon
11	1049 (m)	$r(CH_3)$
12	1153 (w)	$\nu(C-C-O)$ skeletal
13	1299 (m)	$\delta_s(CH_3)$
14	1371 (m.s)	$\delta_a(CH_3)$
15	1444 (s)	$\nu_s(COO^-)$
16	1597 (m)	$\nu_a(COO^-)$
17	1839 (vw)	Al-O (2x) overtone
18	2710 (vw)	
19	2895 (vs)	$\nu(CH)$
20	3041 (w.sh)	$=CH_2$
21	3645 (w.m)	$\nu(OH)$ - surface hydroxyls

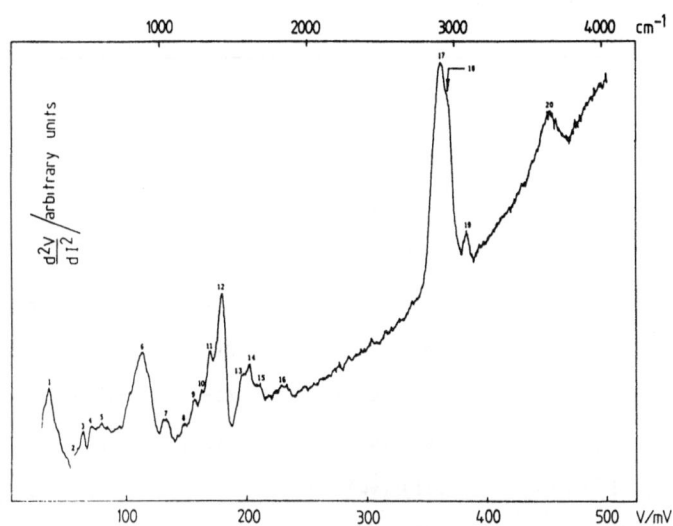

FIGURE 5 — IET spectrum of plasma polymerized acrylic acid

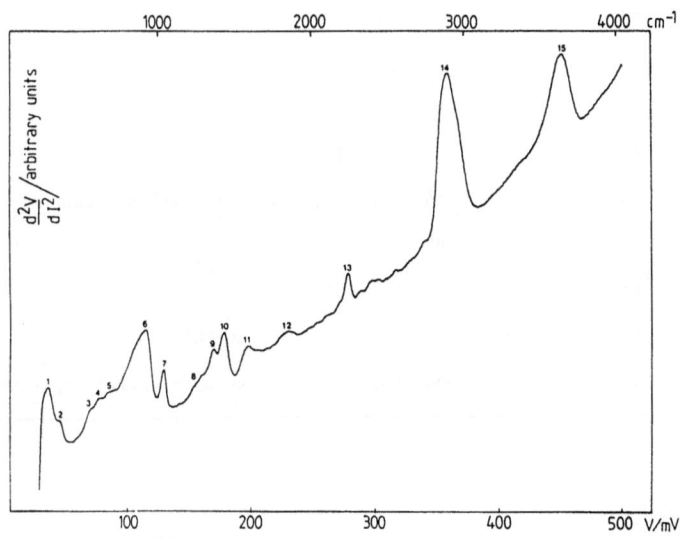

FIGURE 6 — IET spectrum of plasma polymerized acrylonitrile

TABLE 6 - Assignments for plasma polymerized acrylic acid

Peak No.	cm^{-1}	Assignment
1	307 (m)	Al phonon
2	460 (vw.sh)	$r(COO^-)$ in plane
3	500 (w)	
4	581 (w)	$r(COO^-)$ out of plane
5	653 (vw)	$\delta(COO^-)$
6	911 (m)	Al-O bulk phonon
7	1073 (w)	$r(CH_3)$
8	1210 (vw)	
9	1274 (w)	
10	1331 (w)	$\delta_s(CH_3))$
11	1379 (m)	$\delta_a(CH_3)$
12	1460 (s)	$\nu_s(COO^-)$
13	1597 (m.sh)	$\nu_a(COO^-)$
14	1670 (m)	$\nu(C=C)$
15	1710 (m.sh)	$\nu(C=O)$
16	1847 (w)	Al-O (2x) overtone
17	2895 (vs)	$\nu(CH)$
18	2952 (vs.sh)	
19	3049 (w)	$\nu(=CH)$
20	3645 (m)	$\nu(OH)$ - surface hydroxyls

TABLE 7 - Assignments for plasma polymerized acrylonitrile

Peak No.	cm^{-1}	Assignment
1	282 (m)	Al phonon
2	395 (w.sh)	(C–C≡N)
3	557 (w.sh)	
4	621 (w.sh)	OH bend
5	686 (w.sh)	
6	928 (m)	Al-O bulk phonon
7	1049 (wm)	$r(CH_3)$
8	1299 (vw.sh)	$\delta_s(CH_2)$
9	1371 (m)	$\delta_a(CH_2)$
10	1444 (m)	$\nu_s(COO^-)$
11	1605 (wm)	$\nu_a(COO^-)$
12	1855 (w)	Al-O (2x) overtone
13	2234 (m)	$\nu(C≡N))$
14	2879 (s)	$\nu(CH)$ modes
15	3621 (m)	$\nu(OH)$ – surface hydroxyls

Aluminium ions from the substrate would act as counterions, so producing ionic interactions across the interface.   Such interactions are strong and would therefore be beneficial to adhesive joints.   A recent paper by Possart et al. (18) examined the thermal degradation of ultra-thin films of PMMA on steel by mass spectrometry and found methacrylic acid units; these were absent in bulk samples.   It was concluded that water adsorbed at the PMMA-steel interface gives rise to partial hydrolysis of the polymer with methanol as a volatile reaction product.

Using the manner in which the strengths of aluminium-epoxide and aluminium-phenolic adhesive joints change in the presence of wet air as evidence, Comyn et al. (19, 20) proposed that interfacial ion-pairs also occur in these systems.

Elastomer-modified acrylic adhesives are being increasingly used for structural bonding (21).   When aluminium is the substrate, the current evidence would imply that interfacial ion-pairs contribute to bond strength.

4.   CONCLUSIONS

1.   When polymers with ester side groups are adsorbed on alumina there is a high level of ester cleavage to produce a carboxylate anion and an alcohol.   The formation of ion-pairs at the interface is thus an important contribution to the adhesion of acrylic polymers to aluminium.

2.   IET spectra of plasma polymerized VA and MMA are very similar to IET spectra of conventional free radical polymers applied by solution doping.

3.   Transmission IR spectra of plasma polymers and conventionally polymerized PVA or PMMA are very similar.

4.   Polyacrylonitrile is partialy hydrolysed on aluminium oxide.

ACKNOWLEDGEMENTS

The authors gratefully acknowledge the financial support given to the work by the Science and Engineering Research Council, and the valuable discussions held with Professor D.P.Oxley (Leicester Polytechnic) and Dr.P.Poole (Ministry of Defence).

REFERENCES

1. Jaklevic, R.C. and Lambe, R.C.   Phys. Rev. Lett., 17, 1139, 1966.

2. Hansma, P.K.   Phys. Rep., 30(C), 145, 1977.

3. Proceedings of the International Conference and Symposium on Electron
    Tunnelling, (Ed. T.Wolfram), Springer-Verlag, Berlin, 1978.

4. Tegg, J.L., Comyn, J., Horley, C.C., Oxley, P.D. and Pritchard, R.G.
    J. Adhesion, 12, 171, 1981.

5. Reynolds, S., Oxley, D.P. and Pritchard, R.G.   Spectrochim. Acta,
    38A, 103, 1982.

6. Brewis, D.M., Comyn, J., Oxley, D.P., Pritchard, R.G., Reynolds, S.
    and Werrett, C.R.   Surf. Int. Anal., 6, 40, 1984.

7. Mallik, R.R., Pritchard, R.G., Horley, C.C. and Comyn, J.   Polymer,
    26, 551, 1985.

8. Comyn, J., Horley, C.C., Mallik, R.R. and Pritchard, R.G.   Int. J.
    Adhes. Adhes. - in press.

9. Yasuda, H. and Hirotsu, T.   J. Polym. Sci., (Polym. Chem. Ed.), 16,
    743, 1978.

10. Inagaki, N. and Taki, M.   J. Appl. Polym. Sci., 27, 4337, 1982.

11. Clark, D.T. and Abu-Shbak, M.M.   J. Polym. Sci., (Polym. Chem. Ed.),
    22, 1, 1984.

12. Ojha, S.M.   Physics of Thin films.   G.Hass, M.H.Francombe and
    J.L.Vossen (Eds.), 12, 237, 1982.

13. Oxley, D.P., Bowles, A.J., Horley, C.C., Langley, A.J., Pritchard
    R.G. and Tunnicliffe, D.L.   Surf. Interface Anal., 2, 31, 1980.

14. Hall, J.T. and Hansma, P.K.   Surf. Sci., 77, 66, 1978.

15. de Cheveigné, S., Gauthier, S., Klein, J., Léger, A., Guinet, C.,
    Belin, M. and Defourneau, D.   Surf. Sci., 105, 377, 1981.

16. Walmsley, D.G. and Nelson, W.J.   ´Tunnelling Spectroscopy,
    capabilities, applications and new techniques´ (Ed. P.K.Hansma),
    Plenum Press, New York, Ch.11., 1982.

17. Langley, A.J.   Ph.D. Thesis, School of Physics, Leicester
    Polytechnic.

18. Possart, W., Yudin, V.S., Redkov, B.P., Ziegler, H.J., Pozdnyakov,
    O.F. and Bischof, C.   Acta Polymenza, 36, 631, 1985.

19. Comyn, J.   ´Durability of Structural Adhesives´, Ed. A.J.Kinloch,
    Appl. Sci. Publishers, Ch.3, 1983.

20. Comyn, J., Brewis, D.M. and Tredwell, S.T.   J. Adhesion - in press.

21. Lees, W.A.   ´Adhesion´, Ed. K.W.Allen, Elseiver Applied Science
    Publisher, Ch.9, 1984.

Chapter 4

ULTRASONIC EXPLORATION OF ADHESIVE BONDS
BY ACOUSTIC MICROSCOPY

Dominika M. Thaker and Nigel J. Burton

VG Semicon Limited,
Birches Industrial Estate,
Imberhorne Lane,
EAST GRINSTEAD,
West Sussex,
RH19 1XZ

Abstract

The acoustic scanning microscope is a new high resolution
instrument for non-destructive evaluation and testing.

The principles and applications of acoustic microscopy
will be described in detail and the application to adhesive
bonding will be illustrated. The samples investigated include
epoxy bonded aluminium plates with deliberate disbonds, die
attach problems in the semiconductor industry and delamination
studies of carbon fibre/epoxy composites.

Aluminium plates were subjected to various surface
treatments and known polytetrafluoroethylene (PTFE) defects
introduced into the bond layer. These defects, together with
the spacing mesh were resolved and the influence of surface
treatments upon the resulting micrographs discussed.

Unlike other non-destructive testing methods precise
information regarding the location and depth where defects
occur can be obtained.

1.0 Introduction

It may appear, in this day and age, that there are enough tools available for examining materials. However, each have their limitations, the scanning electron microscope can only be used to examine surface structures and not only do the samples have to be able to withstand a vacuum, but also an electron beam which can, and often does, cause damage. The optical microscope cannot examine the interior of opaque materials and X-rays give no information as to exactly where a defect occurs.

Acoustic microscopy offers some considerable advantages over the above named techniques. Not only can information be obtained as to where, say, in a multilayer device a defect occurs, but thickness measurements and mechanical or elastic properties can also be determined. The method is completely non-destructive and non-intrusive: all that is required is that the material should be able to withstand immersion in distilled water.

This article will mainly be concerned with low frequency (45MHz) scanning acoustic microscopy and although high frequencies of around 1GHz are readily available they are of limited application in the investigation of adhesive bonds and bonding in general.

2.0 Principles of the Scanning Acoustic Microscope

The VG Semicon ASM 100 scanning acoustic microscope can be considered to comprise of a lens assembly, scanning system, electronics processing and data acquisition. The most important component is, of course, the lens itself which acts as both the transmitter and receiver of the acoustic pulse.

The lens (Fig 1) is manufactured from a rod of quartz; on one face a spherical surface is ground and situated on the opposing face is a piezoelectric quartz or lithium niobate transducer.

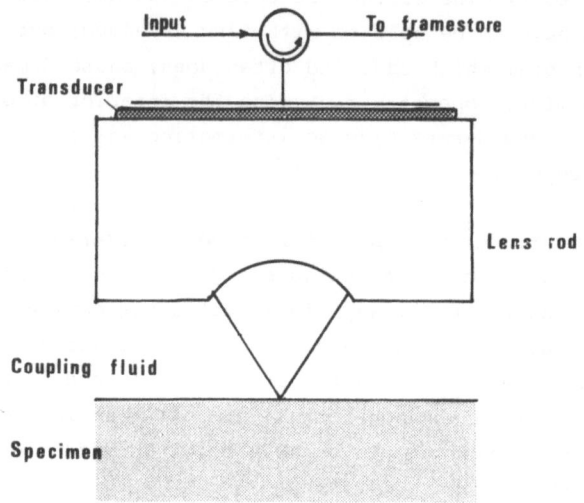

Fig. 1.   Schematic Diagram of the Lens Configuration in a Reflection Acoustic Microscope

When a short electrical pulse is applied a collimated acoustic wave is generated which travels through the rod and is brought to a focus by refraction at the lens surface.

To transmit the acoustic pulse both to and from the sample a coupling fluid is required as the transfer of energy via an air gap is virtually impossible[1]. Water, although not ideal, is most commonly used for this purpose[2].

The reflected echoes are collected by the lens and converted back to an electronic signal which can be displayed on an oscilloscope.

Fig. 2 depicts a typical oscilloscope voltage versus time
trace. The far left-hand peak is the initial output or
transmitted pulse followed by (in the centre) a lens echo,
then a peak reflected from the sample surface and finally one
from inside the specimen. By means of a moveable time-gate
the required pulse can be electronically isolated and its
amplitude at any given point sent to a digital framestore and
used to determine the brightness of a corresponding pixel on a
television monitor. A complete two-dimensional image is then
built up point-by-point by mechanically translating the sample
in an X-Y raster pattern. The magnification is then simply
the ratio of the display size to the scan size and can vary
from X1 to X10,000.

The velocity of sound in water is 1500ms^{-1}, therefore, at
a frequency of 45MHz the surface resoltion is approximately
30μm. Increasing the frequency increases the resolution so

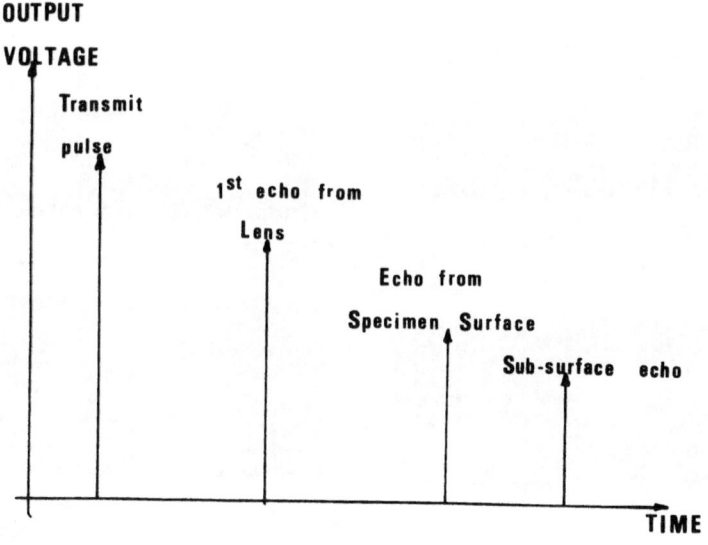

Fig. 2.   Oscilloscope Trace.

that at 2GHz a wavelength of 0.75μm is obtained which is comparable to that of optical microscopy[3]. The smaller wavelengths are not always desirable, however, since as the frequency increases so the depth penetration decreases. At 1GHz only the first 5-10μm of the sample can be viewed, whereas at 45MHz defects have been determined through 6mm of aluminium[4].

The main advantages of the acoustic microscope lies in the ability of ultrasonic waves to penetrate opaque materials. The following examples illustrate how this technique can be used to solve problems occurring in adhesive science.

3.0 Acoustic Microscopy of Packages, Composites and Joints
A consuming problem in the electronics industry is the bonding of semiconductor substrate dies to heatsinks. Continuous thermal cycling leads to costly failures if the die or chip is not adhered correctly.

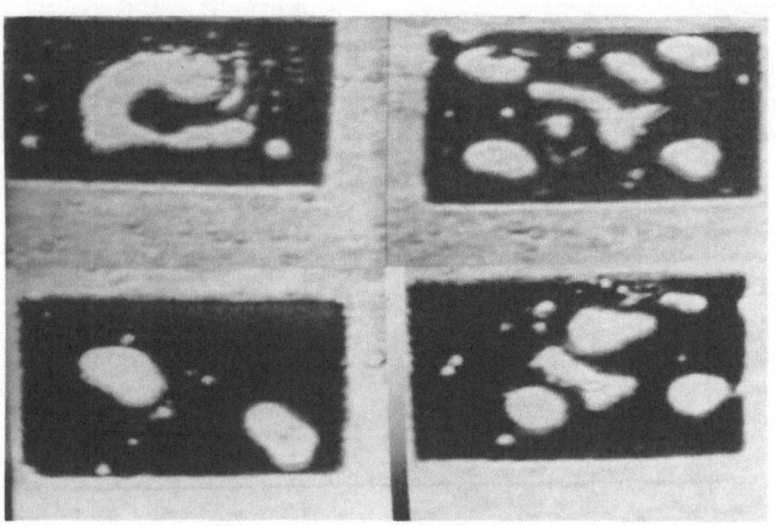

Fig. 3   T0220 Transistor Package Die/Heatsink Bonds

A composite of four die/heatsink bonds in T0220
transistor packages is shown in Fig.3. Some areas of the bond
give a strong reflection and thus appear bright. These
correspond to non-attached regions which, due to their
severity, indicate that the devices are liable to breakdown
and subsequently fail.

Figs. 4 and 5 were imaged through one millimetre of
ceramic and show, respectively, a solder to ceramic substrate
bond and, slightly deeper, the semiconductor die to solder
layer.

Circular voids are plainly visible together with
inhomogeneities at the die/solder interface. Attention
should therefore be given, not only to how the die is
attached, but also, to the method of placing solder onto the
ceramic.

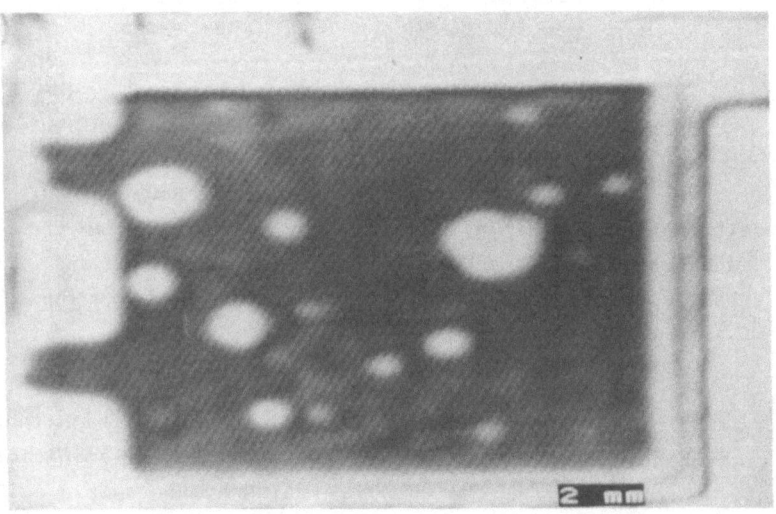

Fig. 4.   Solder Ceramic Substrate Bond on a Hybrid
           Device

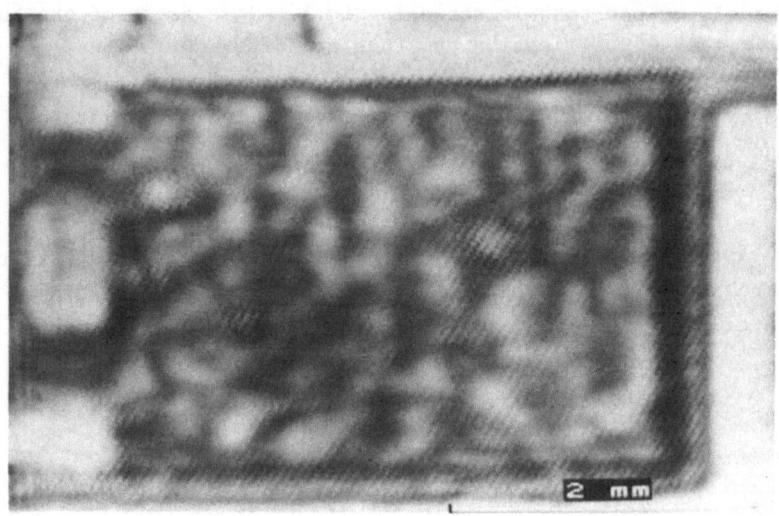

Fig. 5.   Semiconductor Die/Solder Bond on a Hybrid
           Device

Carbon fibre/epoxy resin composites have recently become
popular, in the military and aviation industries, for their
light weight and high strength properties. However,
delaminations within a composite can severely weaken the
transverse strength and lead to failures. Surface and
subsurface images of an impact tested composite are shown in
Figs. 6 and 7.

A small ball bearing had been dropped onto the surface (bottom
right-hand corner in Fig. 6) and the resulting delamination
at, approximately, 0.25mm depth reveals why the sample failed
all strength tests.

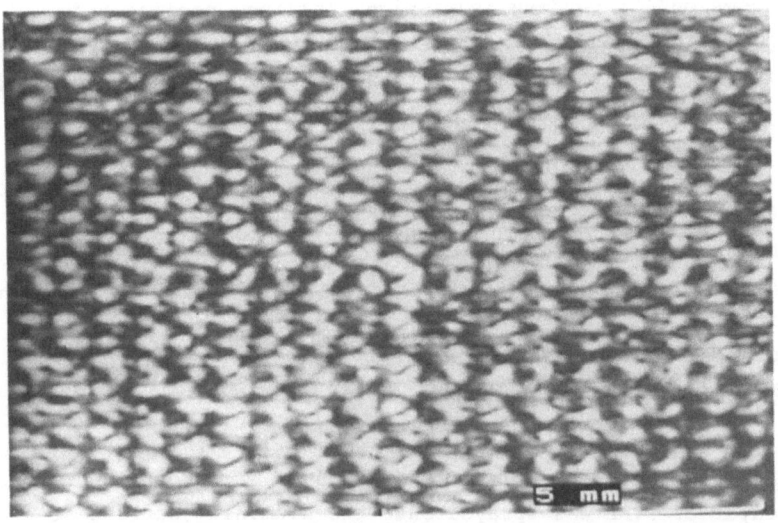

Fig. 6   Surface Image of Carbon Fibre/Resin Composite

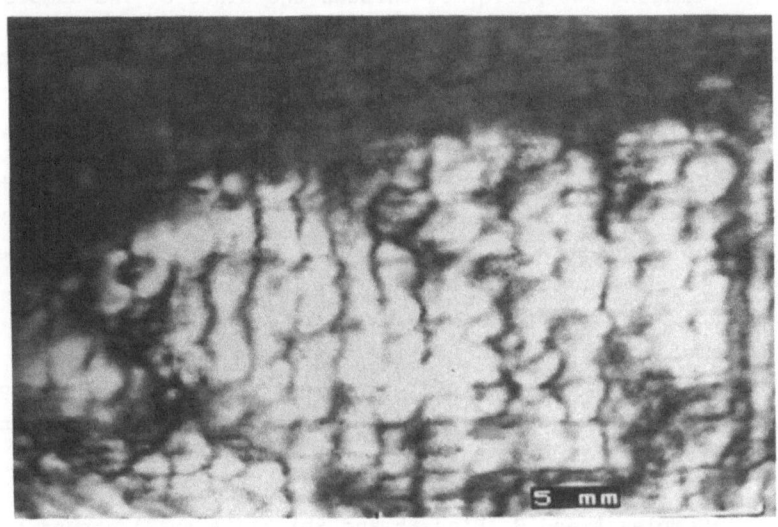

Fig. 7   Subsurface Image of Carbon Fibre/Resin
          Composite

There are numerous techniques used for the measurement of adhesion[5]; bending, squashing, pulling, peeling and hammering among many others. Unfortunately, they are all inherently destructive. Ultrasonic exploration by acoustic microscopy can become a major non-destructive alternative.

Rectangular aluminium alloy bars, each three millimeters thick, underwent different surface preparation techniques before being bonded with a 120°C curing epoxy film adhesive. A spacing mesh was included resulting in a glue line thickness of 0.13mm. Sample A was merely degreased, B - degreased, pickled and acid anodized and sample C degreased and grit blasted.

Adhesive was not applied to one end of each joint and a thin layer of PTFE was sprayed onto the aluminium surface to introduce a small disbond within the joint itself.

Imaging was performed through the sides of the sample in order to obtain a complete picture of the glue line. Figs. 8, 9 and 10 depict the metal to epoxy interface within samples A, B and C respectively.

The large bright areas on the right of each micrograph represents the non-adhered section of each joint. The dark diagonal line seen on Figs. 9 and 10 is due to a deep score mark on the surface of these samples, which displaces both the surface and subsurface peaks, causing them to move out of range of the time-gate. On the left-hand side of the micrographs is the adhered region incorporating the spacing mesh. In places this is visible indicating non-adhesion in that area.

Samples A and B (Figs 8b and 9b) had a small amount of PTFE applied to one surface in the non-adhered (void) section. The PTFE appears dark indicating a bond between the polymer and aluminium alloy. Imaging deeper into these devices one

65

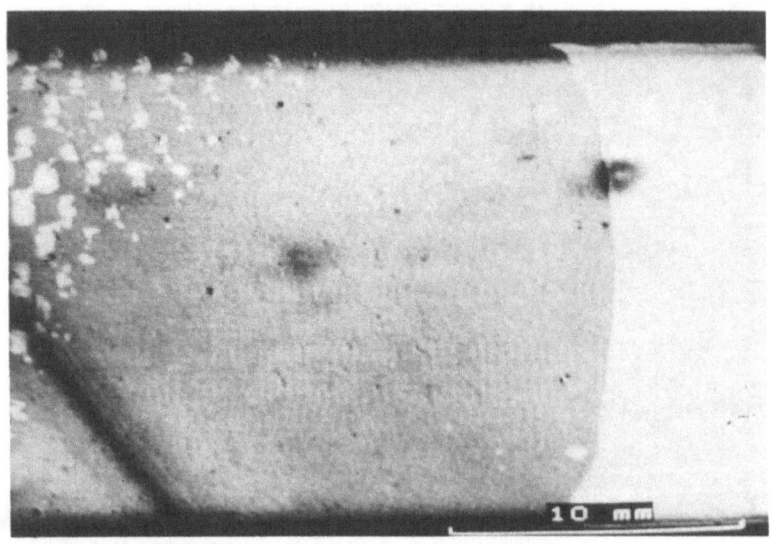

Fig. 8a   Aluminium alloy/epoxy interface.   Sample
           degreased only, side 1

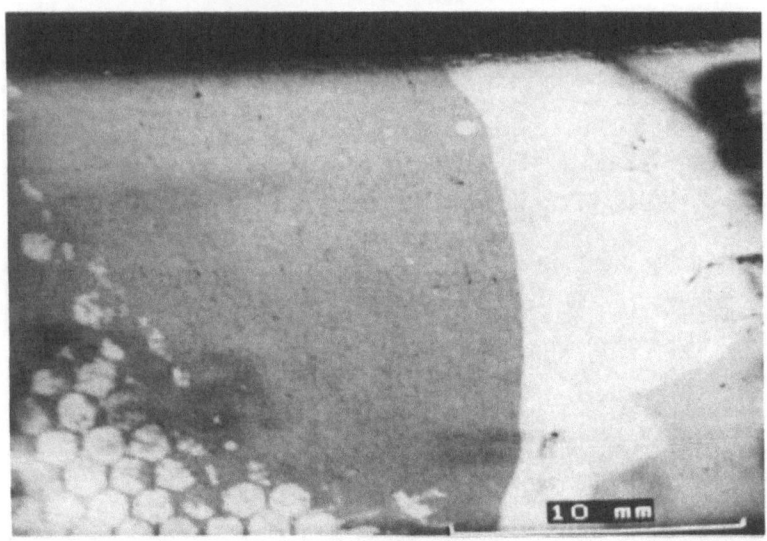

Fig. 8b   Aluminium alloy/epoxy interface.   Sample
           degreased only, side 2

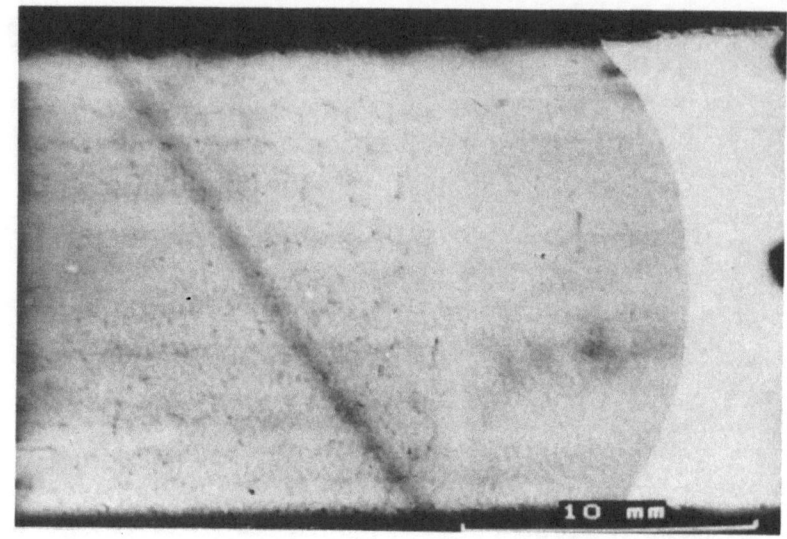

Fig 9a   Sample B, acid treated Al/epoxy interface,
          side 1

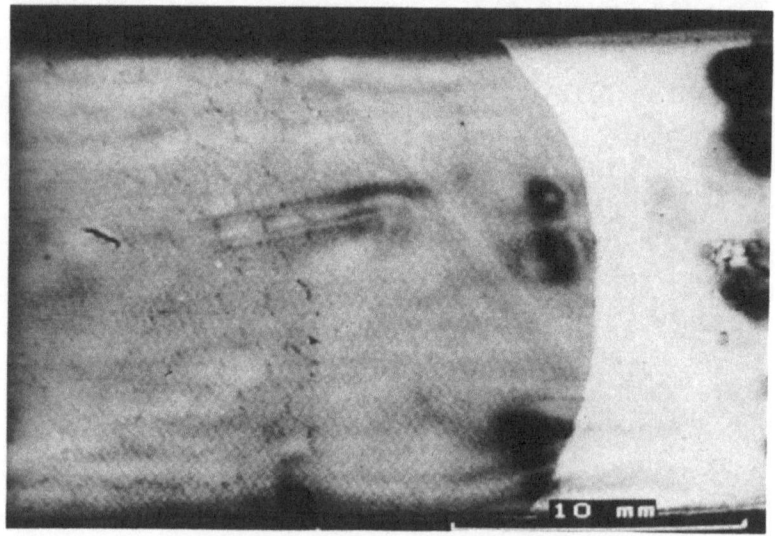

Fig. 9b   Sample B, acid treated Al/epoxy interface,
           side 2

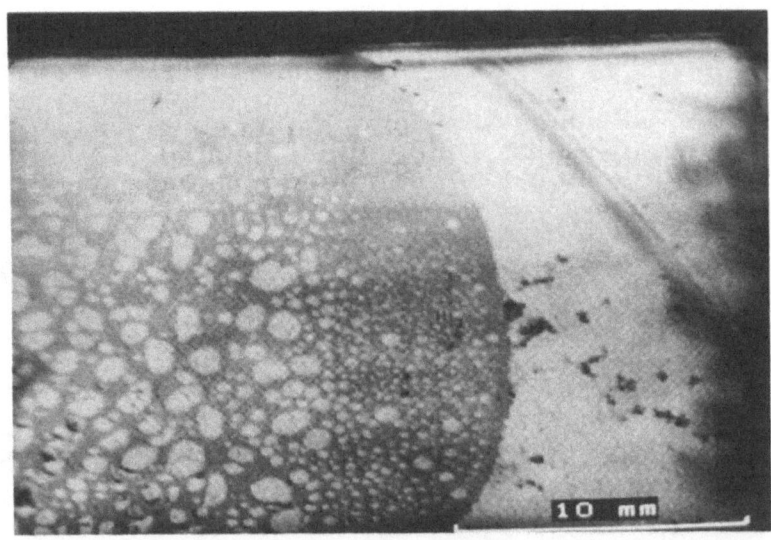

Fig. 10a Sample C, grit blasted Al/epoxy interface,
          side 1

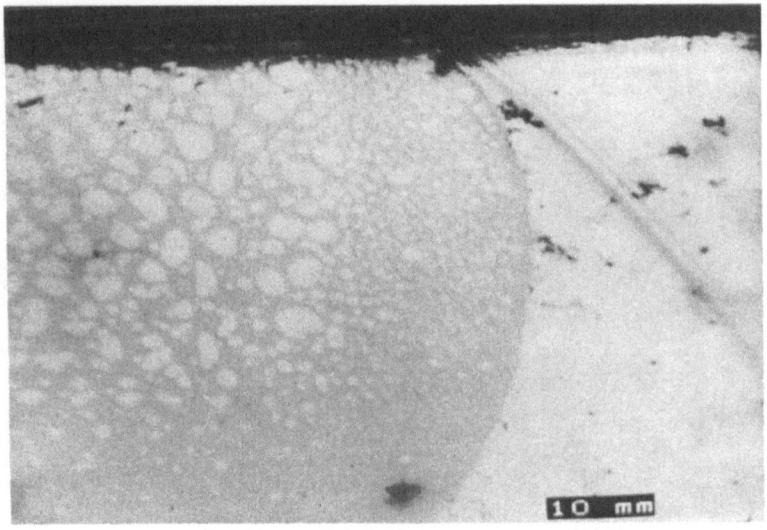

Fig. 10b Sample C, grit blasted Al/epoxy interface,
          side 2

68

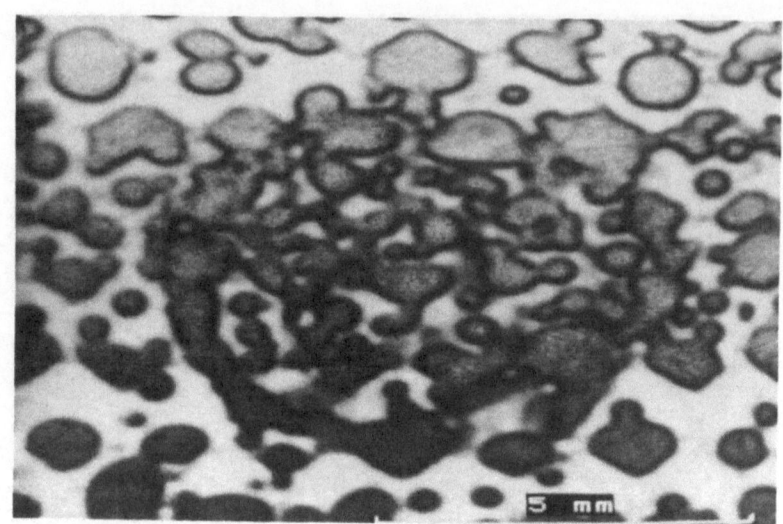

Fig. 11   PTFE defect.   Field of view 10.24mm.

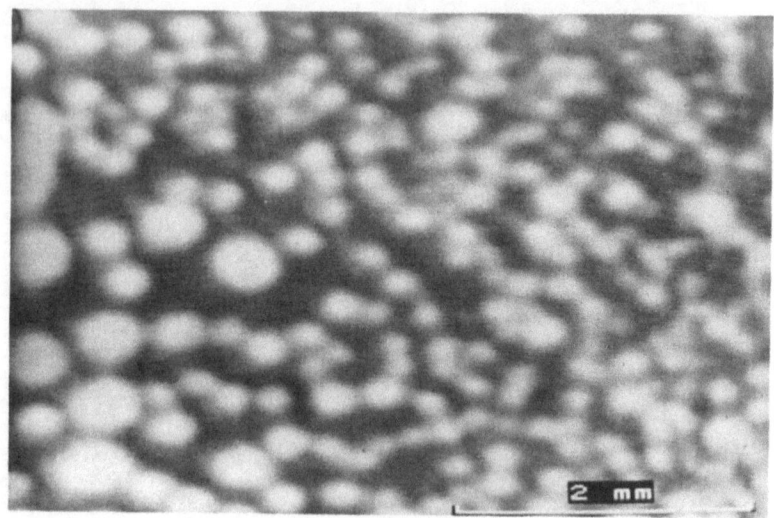

Fig. 12   Detail of adhesive bond.   Grit blasted sample.
Field of view 5.12mm.

would observe the contrast between the non-bonded and PTFE areas to reverse. As the unattached aluminium face completely reflects the acoustic wave a "shadow" would be seen when imaging at greater depth, whereas with the thin polymer layer the acoustic pulse would initially be transmitted, but reflected back when it encounters the void. Thus, at greater depths, the PTFE region could appear bright and the void region dark. PTFE was also used to introduce a circular defect within one of the joints (Fig. 11).

The sample treated by the recommended surface preparation technique of pickling and acid anodizing exhibits, acoustically, an extremely homogeneous bond (Fig.9). The degreased only joint (Fig. 8) appears reasonable, but does not seem as satisfactory as the above. Grit blasting, however, causes the bond to be extremely poor. The internal spacing mesh is not clearly visible (Fig. 10) together with numerous irregular sized "bubbles" (Fig. 12) indicating that this preparation technique is unsatisfactory for the manufacture of such joints.

4.0   Conclusions
A variety of problems have been studied using the VG Semicon ASM 100 scanning acoustic microscope. These studies are of a preliminary nature and further work is needed to relate surface preparation techniques to adhesive quality as seen acoustically. Nevertheless, it is hoped the micrographs here indicate some of the fields for which acoustic microscopy may prove a useful technique.

References

1. M.G. Silk "Ultrasonic Transducers for non-destructive testing". Adam Hilger Ltd., Bristol 1984.

2. C.R. Petts and H.K. Wickramasinghe. Elec Lett. 1980, 16(1), 9.

3. C.F. Quate. et al. Proc IEEE 1979, 67(8) 1092.

4. Work undertaken at VG Semicon.

5. Chapter 4 in "Acoustic and Photoacoustic Microscopy" by R.C. Bray. C.L. Report No. 3243. March 1981. Stanford U.Cal.

Chapter 5

EQUILIBRIUM OF THE TRIPLE LINE SOLID/LIQUID/FLUID OF A SESSILE DROP

M.E.R. SHANAHAN[*] and P.G. DE GENNES[†]

*Centre de Recherches sur la Physico-Chimie des Surfaces Solides, Ecole Nationale Supérieure de Chimie de Mulhouse, Mulhouse Cedex, France.
†Collége de France, Laboratoire de Physique Théorique, Paris Cedex, France.

1   INTRODUCTION

The equilibrium at the triple line where a liquid and a fluid (either vapour or a second liquid immiscible with the first) meet on a solid surface was originally described nearly two centuries ago[1]. By using a simple vectorial argument, the well-known Young equation may be obtained by resolution of the three interfacial tensions, γ, parallel to the solid surface:

$$\gamma_{S2} = \gamma_{S1} + \gamma_{12} \cos \theta \qquad\qquad \ldots \; 1$$

where 1, 2 and S represent respectively the liquid, the fluid and the solid and θ is the contact angle measured in phase 1.   Nevertheless, an objection has on occasion been presented.   Although everything is balanced parallel to the solid surface, nothing would seem to counteract the vertical component $\gamma_{12}\sin \theta$.[2,3]   When the solid is treated as perfectly rigid, it is possible to apply variational calculus and the criterion of minimum free energy at equilibrium.   The result is that equation 1 is perfectly correct.[4-9]   When the solid is considered to be elastic, but very thin, a variational treatment leads us to take into account, in addition to interfacial effects, those due to elastic strain energy and (implicitly) gravity[10].   The approach invokes the modelling of the solid either by thin plate or membrane theory.   This treatment leads to modified equilibrium conditions although in practice the effect will be very small except for very thin solids (cell walls?).   However, the original problem

proposed by Bikerman[2,3] still exists. Local deformation of the solid at
the triple line due essentially to the physical tension γ_{12} is neglected.
The purpose of the present contribution is to examine the immediate neigh-
bourhood of the triple line. Although both Lester[11] and Rusanov[12] have
previously considered this aspect, the present approach is somewhat simpler.

2. STRESS ANALYSIS NEAR THE TRIPLE LINE

The situation near the triple line can be compared to two classic
cases known in stress analysis. We shall consider both of these cases and
show that, to within an additive constant to be discussed later, they lead
to the same result.

2.1 Point Force Applied to a Semi-Infinitive Elastic Solid

Consider a semi-infinitive elastic solid with the origin at the
surface and the distance from the origin along the surface given by ρ
(see Figure 1). The coordinate perpendicular to the surface is represented by z. Landau and Lifshitz[13] show that if a point force, F_z, is applied
at the origin, the corresponding vertical displacement, u_z, is given by:

$$u_z(\rho) = \frac{(1 - \nu^2) F_z}{\pi E \rho} \qquad \qquad \cdot \quad \cdot \quad 2$$

where ν and E are respectively Poisson's ratio and Young's modulus of the
solid.

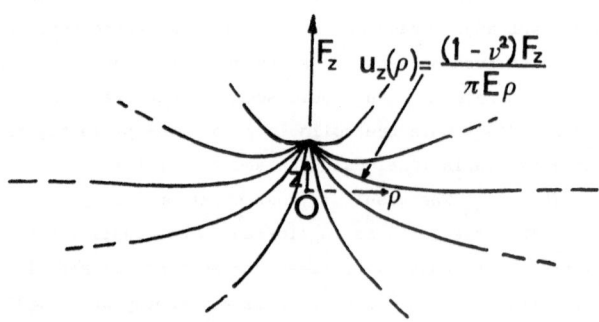

Figure 1: Point force applied to a semi-infinitive elastic solid.

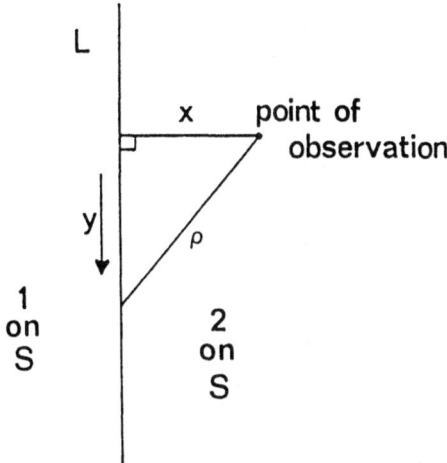

Figure 2:   Plan view near triple line and observation point.

    Now consider Figure 2 which represents a plan view of the region
very near the triple line of a sessile drop on a solid surface. The triple
line itself can be considered to be straight given that the region in question is very small compared to the sessile drop's overall diameter.   In
the following, gravitational effects are taken to be negligible.   Now the
vertical component of surface tension acting on the triple line is $\gamma_{12}\sin\theta$,
where θ is measured with respect to the undeformed solid.   For simplicity
we shall take θ to be $90°$, in which case γ_{S1} and γ_{S2} are equal (γ_{12} is
simply replaced by $\gamma_{12}\sin\theta$ in the following if $\theta = 90°$).   We thus have a
force of γ_{12} per unit length acting vertically on the triple line.
Consider an observation point at a small distance x from the line (this
may, in fact, be on either side of the line).   We can thus evaluate the
vertical displacement at x due to γ_{12} acting on the line by integrating
equation 2 over the "active" length L. F_z is simply replaced by γ_{12}, and
since L >> x, we have equation 3.   Thus we can see that the essential form
of the vertical surface displacement of the semi-infinitive block near the
triple line is $u_z(x) \sim \ln|^{1}\!/\!x|$.   This is shown schematically in Figure 2.
There remains the problem of an unknown constant L and clearly equation 3
cannot be correct at the origin because of the divergence of the integral.
These two aspects are considered below.

$$u_z(x) = \int_{-L/2}^{L/2} \frac{(1-v^2)\,\gamma_{12}}{\pi E\,p} \cdot dy$$

   . . . 3

$$= \frac{(1-v^2)\,\gamma_{12}}{\pi E} \int_{-L/2}^{L/2} \frac{dy}{(x^2+y^2)^{1/2}} \simeq \frac{2(1-v^2)\,\gamma_{12}}{\pi E} \cdot \ln\left|\frac{L}{x}\right|$$

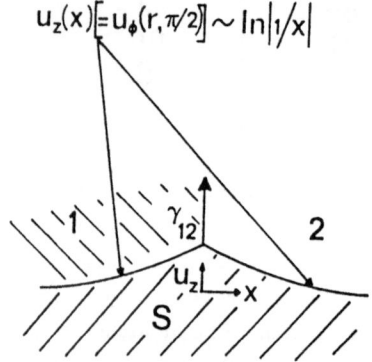

$$u_z(x)\bigl[=u_\phi(r,\pi/2)\bigr] \sim \ln\left|1/x\right|$$

Figure 3:   Schematic representation of deformed solid profile.

2.2   Homogeneous Force Acting on the Edge of a Thin Plate

The second classic case of stress analysis applicable to the present problem is that of a concentrated, homogeneous, vertical force, P, applied to the horizontal edge of a semi-infinitive thin place (unit thickness), as considered by Timoshenko and Goodier.[14]   Figure 4 demonstrates the situation in which the compressional force P is directly replaced in our case by $(-\gamma_{12})$.   This two-dimensional problem can be solved in polar coordinates (r, ϕ) and the Airy stress function, ψ, is:

$$\psi = \frac{\gamma_{12}}{\pi} r\,\phi \sin\phi$$

   . . . 4

Both the tangential stress, σ_ϕ, and the shear stress, $\tau_{r\phi}$, are everywhere zero with this stress distribution.   The remaining radial stress, σ_r, is given by:

$$\sigma_r = \frac{2\gamma_{12} \cos\phi}{\pi r} \qquad \qquad \cdots \quad 5$$

Equation 5 will be of use below for estimating inferior cut-off values for the validity of our treatment. Now the Airy function is valid for the two-dimensional problem whether a state of plane stress or a state of a plane strain is being considered. This is generally the case in the absence of variable body forces[15]. Timoshenko and Goodier consider a thin plate and therefore a state of plane stress since no lateral constraints exist. However, we may consider our "plate" as a "slice" taken from a semi-infinitive block of solid. Lateral deformation is then prevented and we have a state of plane strain (with accompanying lateral stresses due to Poisson's ratio effects). The overall result is that we replace E and ν in the plane stress analysis by $E(1-\nu^2)^{-1}$ and $(1-\nu)^{-1}$ respectively[16]. We may thus derive expressions for the tensile strains, ε_r and ε_ϕ :

$$\varepsilon_r = \frac{\partial u_r}{\partial r} = \frac{2(1-\nu^2)\gamma_{12}\cos\phi}{\pi E r} \qquad \cdots \quad 6$$

$$\varepsilon_\phi = \frac{u_r}{r} + \frac{1}{r}\cdot\frac{\partial u_\phi}{\partial \phi} = \frac{-2\nu(1+\nu)\gamma_{12}\cos\phi}{\pi E r} \qquad \cdots \quad 7$$

$$\frac{1}{r}\cdot\frac{\partial u_r}{\partial \phi} + \frac{\partial u_\phi}{\partial r} - \frac{u_\phi}{r} = 0 \qquad \cdots \quad 8$$

Suitable integration, rearrangement and substitution using equations 6, 7 and 8 leads us to expressions for the radial (u_r) and tangential (u_ϕ) displacements in which three constants are present. Details are to be found in reference 14. By making the reasonable supposition that the solid directly below the origin is not displaced laterally, two constants are eliminated. Now, by considering Figure 4, we may see that the displacement of direct interest is that corresponding to u_ϕ (r, $\pi/2$) = $-u_\phi$ (r, $-\pi/2$). This may be written:

$$u_\phi\left(r, \pi/2\right) = \frac{(1+v)\, \gamma_{12}}{\pi E}\left\{2(1-v)\cdot \ln\left(\frac{d}{r}\right) - 1\right\} \qquad \cdots \quad 9$$

where $\dot{d}$ is a constant and represents the depth at which the solid is no longer deformed vertically by γ_{12}. Comparison of equations 3 and 9 and of Figures 3 and 4 shows the equivalence of $u_z(x)$ and u_ϕ $(r, \pi/2)$. Both equations show a logarithmic dependence on distance (x or r at $\phi = \pi/2$) and the same prefactor. The only difference is the additive constant to be considered below. We shall therefore combine equations 3 and 9 under the form:

$$u_z(x) = \frac{2\left(1-v^2\right)\,\gamma_{12}}{\pi E}\cdot \ln\left|\frac{1}{x}\right| + K \qquad \cdots \quad 10$$

where K is the additive constant.

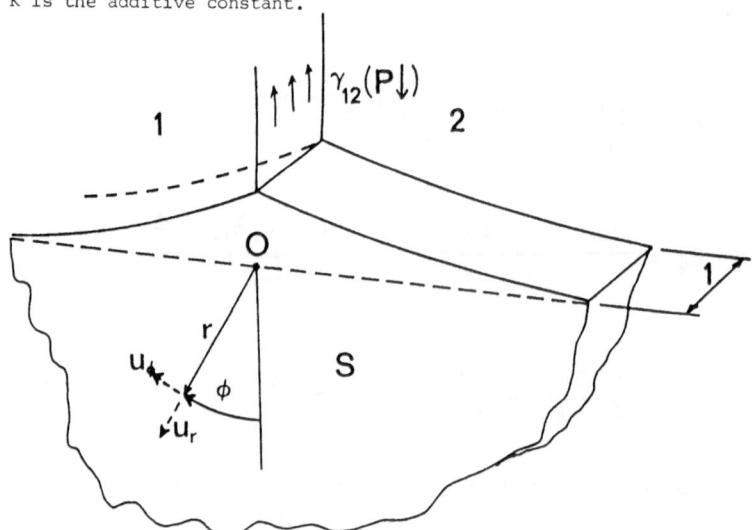

Figure 4:   Vertical force applied to thin plate and corresponding polar coordinates.

3.   LOCAL DEFORMATION OF A RUBBER SURFACE

In the special case of the solid being a rubber, we may exploit equation 10 directly at molecular level. It is known from the theory of rubber elasticity that the modulus may be expressed as[17] :

$$E = \alpha n k T \qquad \cdots \quad 11$$

where n is the number of independent molecular chains per unit volume, kT is the usual kinetic factor composed of the Boltzmann constant and the absolute temperature, and α is a numeric factor of the order of one.   The factor n in turn can be expressed as $(Nv)^{-1}$ where N is equal to the number of monomers between cross-links and V is the volume of a monomer.   For a rubber, Poisson's ratio is close to a half.   Using this value in equation 10, we may calculate the local gradient of the solid surface near the triple line, p(x) :

$$p(x) = \left| \frac{\partial u_z}{\partial x} \right| = \frac{3\gamma_{12}}{2\pi E |x|} \sim \frac{h}{|x|} \qquad \cdots \quad 12$$

where h is the approximate height of the ridge produced by γ_{12}.   Simple rearrangement and use of the rubber elasticity relations leads to:

$$h \sim \frac{3N}{2\pi\alpha} \cdot \frac{\gamma_{12} v}{kT} \qquad \cdots \quad 13$$

Now the order of magnitude of γ_{12} v/kT is that of a monomer diameter, a, and as a result we have the order of magnitude relationship:

$$h \sim N a \qquad \cdots \quad 14$$

The height of the ridge produced by the physical tension γ_{12} pulling on the rubber surface is thus of the same scale as the extended length of a molecular chain between cross-links.   Clearly equation 12 is only reasonably valid for p(x) << 1, or $|x|$ >> h.   Nevertheless, this gives us an idea of the order of magnitude of the local surface deformation caused by γ_{12} and can be compared with the calculation in Section 5.   In the region very close to the triple line, linear elastic theory is no longer applicable anyway due to high stresses (cf. equation 5), and it is thus of little use to apply equation 12 to the region corresponding to $|x|$ < h.   This is treated below.

4.   THE LIMITS OF THE ELASTIC TREATMENT

Clearly as we approach the triple line, problems appear in the elastic theory developed.   It is evident that equation 10 diverges at the origin.   Equivalently, on considering equation 5, we see that for r sufficiently small, the radial stress, σ_r, will reach some cut-off value, σ_c, after which the material will no longer behave elastically, at least in a linear manner.   Given the form of equation 5 (the dependence cos ϕ/r), we may designate a zone near the triple line within the solid which is circular, at least to a first approximation, and inside which the mechanical properties are no longer linearly elastic.   Nevertheless, this non-linear portion has little effect on the behaviour and stress distribution elsewhere.[14]

There is a second point to consider.   We have treated γ_{12} as a force acting on a line up to present.   This is clearly unphysical since γ_{12} must presumably be "spread" more or less over a thickness of liquid, or interphase 1-2.   We shall therefore take γ_{12} to be diffuse over a thickness $2x_o$, which, to a first approximation, we suggest is of the order of the diameter of a molecule of the liquid - say 4 $\overset{o}{A}$ for water.

Having considered these aspects, the cut-off features of equation 10 may differ depending on the type of solid in question.   Here we consider three possibilities:

(a)   For a "hard" solid, the cut-off will be defined either by the atomic structure of the solid, or by the thickness of the liquid/fluid interphase, $2x_o$.   In both cases the elastic theory fails due to the fact that the continuum approach can no longer be valid in the immediate proximity of the triple line.

(b)   For a "soft" solid, there will be a plastic zone near the triple line and the material will behave more or less like an incompressible liquid.   There are two possibilities. One is that at the confluence of the three phases within the plastic zone, equilibrium will be attained locally as with three fluids (Neumann's triangle).   The other eventuality is

a type of "confusion" or molecular mixing of the phases. In
either case, it is probable that the virtually irreversible
phenomena on a highly localised scale will contribute to
wetting hysteresis[18-20].

(c)   In the case of a rubber, described in the previous section,
       there will at first be a zone of non-linear elasticity after
       the cut-off.   Inside this zone, eventually the very high
       stresses will cause all elastic considerations to fail and
       the final situation will again probably be that of three
       fluids meeting.

Analysis of the zones within the cut-offs, considered above, could
prove very complex mathematically and we shall restrict ourselves here to
consideration of the height of the ridge h.

5.   CALCULATION OF RIDGE HEIGHT

Equation 10 gives us the shape of the deformed solid surface near
the triple line but there remains an unknown additive constant, K, which in
fact only determines a vertical translation.   This translation will depend
on boundary conditions removed from the vicinity of the triple line and
involving equilibrium of the whole system drop/fluid/solid.   We shall
therefore simply consider the relative deformation, or displacement,
$\delta u_z(x)$.

Equation 10 gives:

$$\delta u_z(x) = u_z(x_o) - u_z(x) = \frac{2(1-v^2)\,\gamma_{12}}{\pi E} \cdot \ln\left|\frac{x}{x_o}\right| \qquad \cdots \; 15$$

We assume here that the height of the ridge, h, corresponds to
$u_z(x_o)$, i.e. the value at the extremity of the interphase 1-2.   Equation
15 may be used directly to estimate overall ridge height.   Let us take the
example of a drop of water in air ($\gamma_{12} \sim 70$ dynes. cm^{-1}) or a relatively
soft rubber (E $\sim 10^8$ dynes. cm^{-2}).   A membrane which may be deformed significantly by the presence of a water drop will typically have a thickness,
t, $\lesssim 10\mu m$ [10].   We may therefore calculate $\delta u_z(x)$ "far" from the origin,
say at a distance equal to the membrane thickness, 10 μm.   Application of

equation 15 gives δu_z (10 μm) $\sim$ 350 A and thus δu_z (10 μm)/t $\sim 10^{-3}$. Now although the overall conformation of the solid may be altered by the presence of the liquid drop, it is highly unlikely that the surface deformation caused by γ_{12} will exist at distances x > t. We can therefore take δ_{uz} (10 μm) $\sim$ h and the height of the ridge produced by γ_{12} is of order 10^2-10^3 A. (If the reader is uneasy about the slightly arbitrary value of 10 μm taken for x, it will suffice for him to consider the logarithmic form of equation 15 - enormous variations in the said value still leave us quite satisfactorily in the stated range for h). The validity of our approach is confirmed by Saint-Venant's principle and the very small value of δu_z(t)/t. Boundary conditions governing overall drop equilibrium have little importance locally.

Clearly the same approach is applicable to a bulk rubber as well as a membrane. Nevertheless, for a very thin membrane, our interpretation would need to be modified.

Equation 5 allows us to estimate the diameter, D, of the circular non-linear zone. Typically for a rubber, non-linear elastic behaviour occurs at stresses near the value of the modulus ($\sigma_e \sim$ E) and this leads to a value of ca. 50 A for D. Here then the cut-off corresponds to the description of paragraph 4(c).

A similar calculation has been effected for a mica surface which may be considered to be a "hard" solid. The threshold for non-linearity is then a stress of ca. E/100. Young's modulus for mica is ca. 10^{11} dynes cm^{-2} and Poisson's ratio ca. 1/3 (we ignore anisotropy effects to a first approximation). The other relevant values being as above, we obtain h $\sim$ 0.5 A and D $\sim$ 5 A. As a result, we see that near elasticity stays valid virtually everywhere in the triple line region. As a consequence, the cut-off is described as in paragraph 4(a).

6.   CONCLUSION

When a sessile drop (or other liquid conformation) is in contact with a solid surface, the equilibrium at the triple line parallel to the solid surface is adequately explained by Young's equation. However, the vertical component of the liquid/fluid interfacial tension, $\gamma_{12}\sin\theta$, is a true force and must somehow be compensated by deformation of the solid.

This deformation can be calculated using well-known cases of stress analysis. The result is a ridge in the solid obeying a logarithmic law. In the case of a rubber, the ridge height is of the order of magnitude of the extended length of a molecular chain between crosslinks and thus unlikely, under normal circumstances, to exceed 0.1 μm. For a "hard" solid mica, the ridge height is typically so small as to be of atomic dimensions.

REFERENCES

1. Young, T., Phil. Trans. R. Soc. London, 95, 65, 1805.

2. Bikerman, J.J., J. Phys. Chem., 63, 1658, 1959.

3. Bikerman, J.J., 'Physical Surfaces', Academic Press, London, dr.6, 1970.

4. Collins, R.E., Cooke, C.E., Trans. Faraday Soc., 55, 1602, 1959.

5. Johnson, R.E., J. Phys. Chem., 63, 1655, 1959.

6. Goodrich, F.C., 'Surface and Colloid Science', Matijević, E., ed., Wiley, New York, 1, ch.1, 1969.

7. Boruvka, L., Neumann, A.W., J. Phys. Chem., 66, 5464, 1977.

8. Fortes, Phys. Chem. Liq., 9, 285, 1980.

9. Shanahan, M.E.R., 'Adhesion 6', Applied Science Publishers, London, ch.5, 1982.

10. Shanahan, M.E.R., J. Adhesion, 18, 247, 1985.

11. Lester, G.R., Soc. Chem. Ind., Monograph No. 25, London, 57, 1967.

12. Rusanov, A.I., Colloid J. USSR, 37, 614, 1975; English translation of Kolloidn Zh., 37, 687, 1975.

13. Landau, L.D., Lifshitz, E.M., 'Theory of Elasticity', Pergamon, London, p.29, 1959.

14. Timoshenko, S.P., Goodier, J.N., 'Theory of Elasticity', 3rd edition, McGraw-Hill, pp.97-104, 1970.

15. Idem, ibid, p.31.

16. Idem, ibid, p.41.

17. Treloar, L.R.G., 'The Physics of Rubber Elasticity', Clarendon, Oxford, p.66, 1949.

18. Good, R.J., 'Surface and Colloid Science', Good, R.J., and Stromberg, R.R., ed., Plenum Press, New York, 11(1), 13, 1979.

19. Joanny, J.F., de Gennes, P.G., J. Chem. Phys., 81, 552, 1984.

20. Carré, A., Moll, S., Schultz, J., Shanahan, M.E.R., this volume.

Chapter 6

A NOVEL INTERPRETATION OF CONTACT ANGLE HYSTERESIS ON POLYMER SURFACES

A. CARRE[*], S. MOLL[+], J. SCHULTZ[+] and M.E.R. SHANAHAN[+]

[*]Corning Europe Inc., 77211 Avon Cedex, France.
[+]Centre de Recherches sur la Physico-Chimie des Surfaces Solides, Ecole Nationale Supérieure de Chimie de Mulhouse, Mulhouse Cedex, France.

1.   INTRODUCTION

Over the last thirty years or so, the use of contact angle measurement in the study of solid-liquid interactions and problems of adhesion has become very frequent.   Methods have been developed based on the concepts of authors such as Zisman[1] (critical surface tension, γ_c) and Fowkes[2] (polar and apolar interactions), to name but two examples.   The essence of the method of contact angle measurement is that, unless considering a very thin substrate[3], the triple line where the solid, S, liquid 1 and second, immiscible fluid 2 meet can be described by Young's equation relating the three interfacial tensions, γ_{ij}, and the contact angle θ:

$$\gamma_{S2} = \gamma_{S1} + \gamma_{12} \cos \theta \qquad \qquad . \; . \; . \quad 1$$

Provided the surface characteristics of two of the phases are known, those of the third may be assessed using the theoretically unique value of θ. This contact angle will be unique provided the three phases are strictly homogeneous and smooth.   Unfortunately, in practice, it is very common to obtain experimentally a whole range of contact angles for a given three-phase system.   The largest angle obtainable corresponds to that observed just after a drop of liquid has advanced on the solid surface and is therefore known as the advancing angle, θ_A.   Similarly the smallest angle is obtained just after the arrest of a receding liquid drop and the corresponding angle is θ_R.   It is generally assumed that the equilibrium contact angle is to be found near the centre of this range.   This variability, or hysteresis has been attributed to several causes[4], in particular, inhomogeneity of the solid surface[5-7] and its rugosity[8-10]. Other causes noted are diffusion, swelling and reorientation[4].

In many cases of contact angle measurement, the second immiscible fluid phase, 2, is simply air, or more correctly, the vapour of liquid 1. However, in recent years, more has been done in our laboratories on the determination of the surface and interfacial tension characteristics of solids using a two-phase liquid system[11-14]. This method has the advantage of allowing finite contact angles to be obtained for solids on which most normal liquids would spread spontaneously in the absence of the second liquid phase. Nevertheless, the same problem of contact angle hysteresis exists. In a fairly recent study[15], some of us considered the thermodynamics of reversibility of wetting in such two-phase liquid systems and, amongst other things, came to the conclusion that the two contact angles θ_1 and θ_2, i.e. those referring to liquid 1 on the solid in the presence of liquid 2 and liquid 2 on the solid in the presence of liquid 1, should be supplementary. This was borne out experimentally using both polyethylene (PE) and highly-oriented pyrolytic graphite (H.O.P.G.) as model solids. However, it has since been noticed with other solids that the supplementary relationship does not always hold good and this seems to be related to the hysteresis phenomenon. Both PE and H.O.P.G. are essentially apolar solids. The purpose of the present contribution is to consider the rôle of polarity in wetting hysteresis, with particular reference to polymers.

2.   EXPERIMENTAL

Figure 1 represents a sessile drop of liquid 1 on a solid surface, denoted by S, and in the presence of a second, immiscible liquid 2. It is assumed that the surface tension, γ_i, of the solid and of the two liquids is composed of two additive terms, γ_i^D, the dispersive component, and γ_i^P, the polar component. Other interactions, such as those of an acid-base, or ionic nature, for example, are assumed not to intervene. If liquid 2 is apolar, the relationships of Young (above) and Fowkes[16] can be applied and the following equation obtained[11]:

$$\gamma_1 - \gamma_2 + \gamma_{12} \cos \theta_1 = 2 \left(\gamma_S^D \right)^{1/2} \left[\left(\gamma_1^D \right)^{1/2} - \left(\gamma_2^D \right)^{1/2} \right] + I$$

$$\cdots \quad 2$$

where I represents the polar interaction between the solid and liquid 1.
Its nature will be left unspecified here since there is still much doubt as
to its form. Nevertheless, clearly it will be an increasing function of
the polar nature of both the solid and liquid 1. In the special case when
γ_1^D and γ_2 are virtually equal, equation 2 simplifies to:

$$\gamma_1 - \gamma_2 + \gamma_{12} \cos \theta_1 \simeq I \qquad \ldots \quad 3$$

This is the basis and novelty of the method employed in the present context.
If the three γ terms are known and the contact angle measured, a direct
estimation of the polar interaction is accessible. There is no need for
subsidiary experiments to assess polarity. Both contact angle hysteresis
and polarity are directly obtainable. The liquids in question for the
present study are water (W) for which $\gamma_1 = 72.6$ mJ.m^{-2}, $\gamma_1^D = 21.6$ mJ.m^{-2}
and n-octane (H) for which $\gamma_2 = 21.3$ mJ.m^{-2}. $\gamma_{12} = \gamma_{WH}$ is taken to be
51 mJ.m^{-2}

Figure 1:   Sessile drop of liquid 1 on solid surface S in presence of
             fluid 2.

Now by inversion of the system, the contact angle of n-octane in
water on the solid, θ_2, can be measured and the equivalent equation to 3 is
simply:

$$\gamma_1 - \gamma_2 - \gamma_{12} \cos \theta_2 \simeq I \qquad \ldots \quad 4$$

$$\gamma_W - \gamma_H - \gamma_{WH} \cos \theta_2 \simeq I \qquad \ldots \quad 4a$$

The two systems are shown schematically in Fig. 2, the drop being under-
neath the solid in the second case because of the relative densities of
water and n-octane. Subtraction of equation 4 from equation 3 leads to
the supplementarity relationship of angles θ_1 and θ_2:

$$\cos \theta_1 + \cos \theta_2 = 0 \quad \Longleftrightarrow \quad \theta_1 + \theta_2 = 180° \quad \ldots \quad 5$$

In addition, for each system both advancing, θ_A, and receding, θ_R, contact angles may be obtained by either adding to or taking liquid from the sessile drop (or trapped bubble depending on the case), with a microsyringe. Since θ_A corresponds to water encroaching on solid previously in contact with n-octane in case 1 and this is equivalent to θ_R for case 2, the two should be equivalent. We should thus have:

$$\theta_{1A} = 180° - \theta_{2R} \qquad \ldots \quad 6$$

and by an exactly analogous argument:

$$\theta_{1R} = 180° - \theta_{2A} \qquad \ldots \quad 7$$

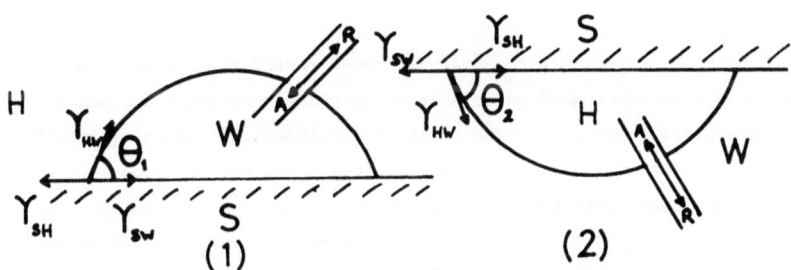

Figure 2:   Contact angles of water in n-octane (1) and n-octane in water (2).   Method of measuring θ_A and θ_R.

Consider the advancing and receding cases of water in n-octane. Assuming that there is hysteresis, I, will refer to the estimated polar interaction using θ_{1A} and I* for θ_{1R}:

$$\gamma_W - \gamma_H + \gamma_{WH} \cos \theta_{1A} \simeq I \qquad \ldots \quad 8$$

$$\gamma_W - \gamma_H + \gamma_{WH} \cos \theta_{1R} \simeq I^* \qquad \ldots \quad 9$$

Subtracting equation 8 from equation 9, the hysteresis of wetting, H, may be defined:

$$H = \cos\theta_{1R} - \cos\theta_{1A} \simeq \gamma_{WH}^{-1}\left(I^{*} - I\right) \quad \cdot \quad \cdot \quad \cdot \quad 10$$

Using equations 6 and 7, an analogous expression is obtained for n-octane in water. The reason for obtaining equation 10 is that it shows how hysteresis can be related directly to a difference in polar interaction energies between the solid and liquid 1.

This then represents the basis of the experimental method and thus the direct measurement of contact angles θ_{1A}, θ_{1R}, θ_{2A} and θ_{2R} was effected on a Rame Hart A-100 contact angle goniometer at room temperature on a series of polymers known in some cases for their relatively apolar nature and in others for their marked polarity. They are: polytetrafluoroethylene (PTFE), polyethylene (PE), polypropylene (PP), polystyrene (PS), polyvinylidene fluoride (PVDF), polyamide 11 (PA11), polyethylene terephthalate (PET), and polymethylmethacrylate (PMMA).

3.   RESULTS

Table 1 represents a summary of experimental results obtained for the various polymers studied using the two-phase n-octane/water and water/n-octane liquid systems. The first four columns of numbers represent average contact angle measurements, each being the mean of about twenty readings and the scatter being $\pm 3^{\circ}$. Given the preceding arguments, the sums $(\theta_{1A} + \theta_{2R})$ and $(\theta_{1R} + \theta_{2A})$ should always be 180°. It can be seen from the fifth and sixth columns of numbers that prediction is reasonably well corroborated by the experimental values (with perhaps the exception of polypropylene (PP)). We have thus found the probable cause of lack of supplementarity of angles θ_1 and θ_2 referred to at the end of the introduction. Unless care is taken, each experimental value will be near its own value of θ_A and if hysteresis exists, in general, $\theta_{1A} + \theta_{2A} \neq 180^{\circ}$. It is therefore suggested that if inversion of a two-phase liquid system in contact angle measurement leads to lack of supplementarity, then hysteresis, as described more fully below, should be suspected.

Since this first part of the study concerning supplementarity is borne out by experiment, we may simplify the following by assuming a direct equivalence between θ_{1A} and $(180^{\circ} - \theta_{2R})$ on one hand, and θ_{1R} and $(180^{\circ} - \theta_{2A})$ on the other.

We shall therefore define for each polymer:

$$\theta = \left[\theta_{1A} + \left(180° - \theta_{2R}\right)\right]\Big/2$$

$$\theta^* = \left[\theta_{1R} + \left(180° - \theta_{2A}\right)\right]\Big/2$$

and to simplify the following, consider that each contact angle corresponds to the system 1, i.e. water in n-octane. In the manner of equations 8, 9 and 10, we can now write:

$$\gamma_W - \gamma_H + \gamma_{WH} \cos\theta \simeq I \qquad . \quad . \quad . \quad 11$$

$$\gamma_W - \gamma_H + \gamma_{WH} \cos\theta^* \simeq I^* \qquad . \quad . \quad . \quad 12$$

$$H = \cos\theta^* - \cos\theta \simeq \gamma_{WH}^{-1}\left(I^* - I\right) \qquad . \quad . \quad . \quad 13$$

where the first relationship applies to water encroaching on the solid previously in contact with n-octane and the second to n-octane encroaching on the solid previously in contact with water, the contact angle being measured in the water phase for both cases.

Table 1:

Polymer	θ_{1A}°	θ_{1R}°	θ_{2A}°	θ_{2R}°	$(\theta_{1A} + \theta_{2R})^{\circ}$	$(\theta_{1R} + \theta_{2A})^{\circ}$
PTFE	174	166	12	0	174	178
PE	171	160	19	0	171	179
PP	171	169	36	0	171	205
PS	168	142	38	0	168	180
PVDF	118	92	87	55	173	179
PA11	121	82	89	55	176	171
PET	120	81	98	60	180	179
PMMA	127	75	103	60	187	178

4.   INTERPRETATION AND DISCUSSION

Table 2 gives a summary of values of θ, θ^*, the hysteresis, H, and the polar interactions water/ solid, I and I^* as calculated from equations

11, 12 and 13. What is clear even on cursory inspection is that the hysteresis H increases with polarity either I or I*. We shall attempt to explain this and attach a semi-quantitative interpretation to it.

Table 2:

Polymer	$\theta°$	$\theta^{*°}$	$H = (\cos \theta^* - \cos \theta)$	$I (mJ.m^{-2})$	$I^* (mJ.m^{-2})$
PTFE	177	167	0.02	0.4	1.6
PE	176	161	0.05	0.4	3.1
PP	176	157	0.08	0.4	4.4
PS	174	142	0.21	0.6	11.1
PVDF	122	93	0.48	24.3	48.6
PA11	123	87	0.60	23.5	54.0
PET	120	82	0.64	25.8	58.4
PMMA	124	76	0.80	22.8	63.6

Let us first assume as a hypothesis that hysteresis is directly related to the orientation of polar groups attached to the macromolecular carbon-carbon chains near the polymer surface in contact with water and n-octane. It has already been established that surface properties of poly-mers can alter depending on the phase in contact and that this is due to molecular mobility of some form or other[17-21]. As a consequence, we shall consider that the values of I and I* correspond to two distinct states of orientation of the surface polar groups, without affecting the overall status of the backbone chains. I represents a random distribution and I* represents a more or less oriented distribution such that polar groups tend to be directed towards the liquid/solid interface. This tendency to orient or not will be a consequence of the system trying to minimise its overall free energy. A simplistic model of the situation is shown in Fig. 3 in which the distribution of polar groups is random near the inter-face with an apolar liquid in contact due to lack of attraction, whereas there is an attraction between the polar groups and the liquid when this latter is of a polar nature, and orientation results. We can thus consider, in the simplest terms, that two types of surface exist -

S representing the random polar group distribution, and S* representing the
ordered distribution.   The intrinsic polar nature of the two will in
general be different.

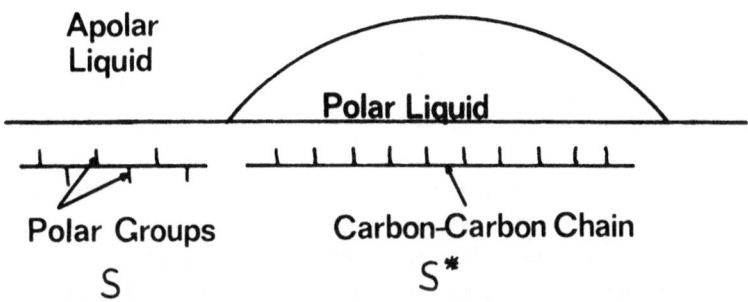

Figure 3:   Schematic representation of polar group orientation in the
                 polymer due to polar liquid contact.

        We shall now make a second assumption.   It is that the kinetics of
orientation or disorientation of the polar groups near the polymer surface
corresponds to a much longer time scale than that associated with the speed
of flow of a liquid drop advancing or receding on the solid during contact
angle measurements.   This means that an advancing drop of water will find
itself in contact with the polymer surface just previously in contact with
n-octane and therefore equivalent to surface S.   The triple line is therefore on solid S.   Inversely, when the drop of water recedes, the contact
line is on the solid previously in contact with water and therefore, by
hypothesis, oriented and represented by S^*.   The situation is represented
schematically in Fig. 4.

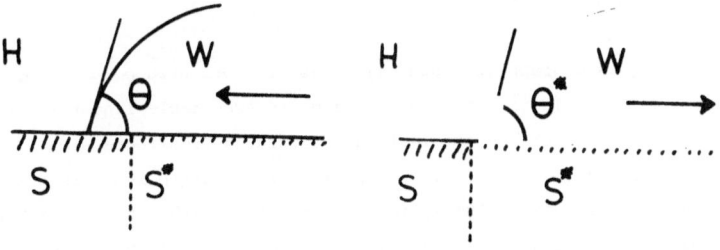

Figure 4:   Hysteresis, H, vs. polarity, I^*, for various polymers.

The ability of polar groups on the polymer chain to orient themselves towards the surface will be controlled by essentially two things:

- the general molecular mobility of the polymer, or more
  specifically the facility of rotation of polar groups
  about the carbon-carbon chain,

- and the intrinsic polarity of the polymer groups and
  that of the contacting liquid, although this latter is
  a constant in the present context.

The first of these may be considered as the relative resistance to
orientation, and the second, the factor encouraging orientation. Clearly,
following the initial hypothesis, if a polymer is polar, but molecularly
"rigid", hysteresis H will be non-existent. Similarly if the polymer is
"flexible" on a molecular level, but totally apolar, H will also be absent.

We shall use the ratio I^*/I to refer to the facility of orientation.
If the ratio is 1, there is no hysteresis irrespective of the actual value
of polarity; if > 1, some orientation is possible. As far as the second
factor is concerned we have the choice of either I or I^* to assess polarity.
However since the initial process of orientation must be triggered by the
already existent polarity, corresponding to the random polar group distribution, or surface S, then it is felt that I is a better measure of the
parameter favouring orientation. We can therefore combine the two influences and suggest that the overall "orientability", O, is given semi-
quantitatively by:

$$O = \frac{I^*}{I} \times I = I^* \qquad \qquad \cdot \ \cdot \ \cdot \qquad 14$$

Hysteresis is then a function of I^*.

Figure 5 is a graph of hysteresis H vs. the oriented polar interaction energy I^*. It can be seen that a quite acceptable linear relationship exists of which the gradient, as calculated from regression
analysis, is 1.14×10^{-2} $mJ^{-1}.m^2$. There is however one polymer in the
series studied which gave slightly doubtful results, although this is more
evident from Table 1 than from Fig. 5. The supplementarity of contact
angles leaves something to be desired in the case of PP. This we do not

understand but it could quite possibly be due to some extraneous effect such as swelling by the alkane.   Nevertheless, the same sort of error could then perhaps be expected with PE but this was not observed.

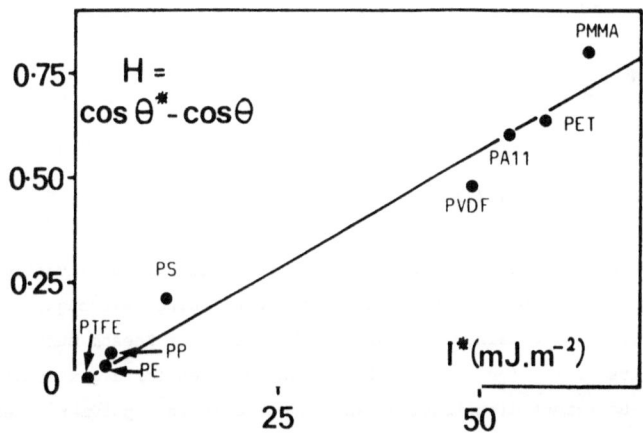

Figure 5:   Hysteresis, H, vs. polarity, I^*, for various polymers.

Having established the very clear correlation between polar inter-action and hysteresis in polymer wetting, it should not be overlooked that effects of orientation have been recognised for some considerable time. The dipole-dipole theory of Keesom was put forward many years ago[22,23]. We have in fact attempted to explain the present results using Keesom's approach but with little success.   The essential assumptions were that the minimum dipole-dipole interaction energy corresponds to our orientated state and is therefore written U^*; and that the average dipole-dipole interaction energy, U, corresponds to our random state.   Under these circumstances,

$$U^* = \frac{-2 p_1 p_2}{\ell^3} \qquad \ldots \quad 15$$

$$U = \frac{-2 p_1^2 p_2^2}{3kT\ell^6} \qquad \ldots \quad 16$$

where p_1 and p_2 and electric dipole moments and ℓ the distance between them[24]. We assume that polar interaction energy and dipole-dipole interaction energy are directly proportional:

$$I \propto U \qquad \qquad \cdot \quad \cdot \quad \cdot \quad 17$$

$$I^* \propto U^* \qquad \qquad \cdot \quad \cdot \quad 18$$

and thus from equations 15 and 16:

$$I^* = c\,I^{1/2} \qquad \qquad \cdot \quad \cdot \quad 19$$

where c is a constant. The parabolic relationship was investigated using the results of Table 2 but found to be unconvincing, although a qualitative correlation was obtained. However, the theory of Keesom must be taken within its limitations. Equation 16 was derived for a gas assuming a Maxwell-Boltzmann distribution, whereas clearly in a polymer, stearic hindrance will prevent free orientation. Nevertheless, further investigation in this line would be of interest and an idea for a simple model which could perhaps be refined, is presented in the appendix to this study.

5.   CONCLUSIONS

The hysteresis of wetting has been recognised for a considerable time and effects such as chemical surface inhomogeneity and surface rugosity of the solid have been accepted as being important causes. Nevertheless this same phenomenon is often observed on polymer surfaces which are both reasonably homogeneous and smooth. Clearly another cause is in question. After having noticed a qualitative correlation between hysteresis and surface polarity on polymers, a quantitative study has been undertaken. This was conducted using the novel approach of employing a two-phase liquid system in which one liquid is apolar and the two have very similar dispersive components to their surface tension. The advantage is that hysteresis and solid polarity can be assessed simultaneously.

Undeniably there is a clear correlation between the polar nature of a polymer surface and wetting hysteresis. A simple model is proposed to explain the phenomenon although this could no doubt be improved upon considerably in the future.

6. APPENDIX: SIMPLE MODEL OF POLYMER ORIENTATION

Considering Fig. 3, we assume as a very simple model that a typical carbon-carbon chain runs parallel to the solid/liquid interface such that any polar groups are free to rotate. A random orientation will be adopted when the contacting liquid is apolar and an ordered one when it is polar. Clearly in practice, few chains, if any, will lie truly parallel to the interface but this does not change the argument since we may consider that most chains near the polymer surface will have segments or sections which are parallel.

Consider now schematically one such polar group on a carbon-carbon chain (Fig. 6). The group is at A, a distance r from the main chain, denoted by C, and able to rotate about it. Now take a point at B at a distance ℓ from the main chain and assumed to be in the liquid above the solid surface. By analogy with electrostatic theory, we may calculate a potential, V, as being a constant K divided by the distance from the polar group:

$$V = K/AB \qquad \cdots \quad 20$$

If all such groups are oriented towards the solid/liquid interface as shown in Fig. 6(a), then the mean potential will be given simply by:

$$V^* = K/(\ell-r) \qquad \cdots \quad 21$$

This corresponds to the oriented polymer in contact with water. However, in the case of a random distribution, as shown in Fig. 6(b), the mean potential must be calculated by averaging over all the possible positions of the polar group, i.e. over the possible values of ϕ, $0 < \phi < \pi$. In this case the mean potential is given by:

$$V = \frac{K}{2\pi} \int_0^{2\pi} \frac{d\phi}{(\ell^2 + r^2 - 2\ell r.\cos\phi)^{1/2}} \qquad \cdots \quad 22$$

Since potentials can be associated with interaction energies, we then assume that:

$$\frac{I}{I^*} = \frac{V}{V^*} = \frac{(\ell - r)}{2\pi} \int_0^{2\pi} \frac{d\phi}{(\ell^2 + r^2 - 2\ell r \cos\phi)^{1/2}} \qquad . . \quad 23$$

Now I and I* are known experimentally and their ratio can be assimilated to a function of r and ℓ. If r is known, an estimate of ℓ can be made to get an idea of the distance between a surface carbon-carbon chain and the polar molecule with which it interacts in the liquid. Unfortunately, it is not too clear what value of r should be taken for many polymers but we may take as a fairly good example that of polystyrene. We take r to be the distance between the carbon-carbon chain and the centre of the benzene ring, i.e. ca. $\overset{\circ}{5}$ Å. Use of equation 23 leads to a value of ℓ of ca. 3.2 Å. This implies that the polar interaction between oriented polymer groups and water takes place essentially with the molecules closest to the interface.

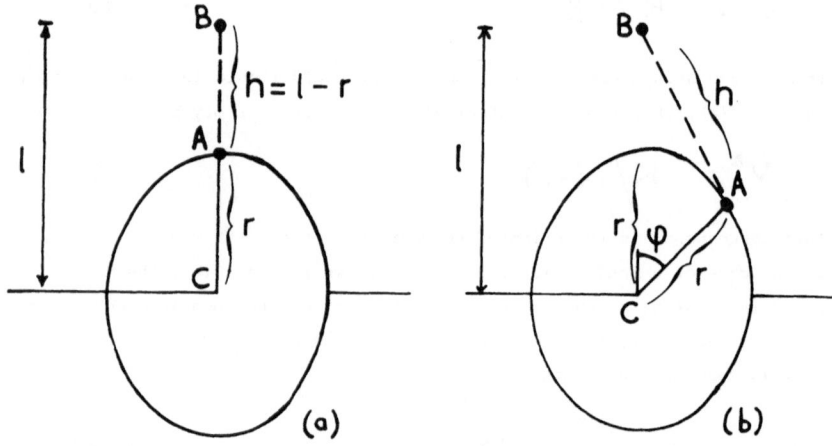

Figure 6:   Simple model of polar group orientation

Clearly this model should not be taken as being too realistic. Apart from the oversimplification of treating the polymer chain or segments thereof as being entirely parallel to the surface, the calculation only takes account of interactions with nearest neighbours. Much improvement is possible to this model, including perhaps the adoption of an interaction potential other than the very simple electrostatic example. Nevertheless, this simple estimate gives a plausible result and the order of magnitude of distance ℓ seems reasonable.

REFERENCES

1.   Zisman, W.A., 'Contact Angle, Wettability and Adhesion', Advances in Chemistry Series, 43, Am. Chem. Soc., Washington D.C., 1, 1969.

2.   Fowkes, F.M., Ind. Eng. Chem., 56, 40, 1964.

3.   Shanahan, M.E.R., J. Adhesion, 18, 247, 1985.

4.   Good, R.J., 'Surface and Colloid Science', Good, R.J. and Stromberg, R.R., ed., Plenum Press, New York, 11(1), 13, 1979.

5.   Cassie, A.B.D., Discussion, Faraday Soc., 3, 11, 1948.

6.   Neumann, A.W., Good, R.J., J. Colloid Interface Sci., 38, 341, 1972.

7.   Penn, L.S., Miller, B., J. Colloid Interface Sci., 78, 238, 1980.

8.   Wenzel, R.N., Ind. Eng. Chem., 28, 988, 1936.

9.   Johnson, R.E., Dettre, R.H., ref. 1, pg. 112.

10.   Eick, J.D., Good, R.J., Neumann, A.W., J. Colloid Interface Sci., 53, 235, 1975.

11.   Schultz, J., Tsutsumi, K., Donnet, J.B., J. Colloid Interface Sci., 59, 272 and 277, 1977.

12.   Carré, A., Schultz, J., J. Adhesion, 15, 151, 1983.

13.   Schultz, J., Simon, H., Vernes et Réfractaires, 34, 23 and 192, 1980.

14.   Shanahan, M.E.R., Cazeneuve, C., Donnet, J.B., Schultz, J., International Conference: Adhesion and Adhesives - Science, Technology and Applications, Durham, England, 19, 1, 1980.

15.   Shanahan, M.E.R., Cazeneuve, C., Carré, A., Schultz, J., J. Chim. Phys., 79, 241, 1982.

16.   Fowkes, F.M., 'Treatise on Adhesion and Adhesives', Patrick, R.L., ed., Marcel Dekker, New York, 1(9), 344, 1967.

17.   Andrade, J.D., Ma, F.N., King, R.N., Gregonis, D.E., J. Colloid Interface Sci., 72, 488, 1979.

18.   Yasuda, H., Sharma, A.K., Yasuda, T., J. Polym. Sci., Polym. Phys., 19, 1285, 1981.

19.   Carré, A., Schreiber, H.P., J. Coatings Technol., 54, 31, 1982.

20.   Schultz, J., Carré, A., Mazeau, C., Int. J. of Adhesion and Adhesives, 4, 163, 1984.

21.   Lavielle, L., Schultz, J., J. Colloid Interface Sci., 106, 438, 1985.

22.   Keesom, W.H., Physik. Z., _22_, 126, 1921.

23.   Keesom, W.H., Physik, Z., _23_, 225, 1922.

24.   Good, R.J., 'Treatise on Adhesion and Adhesives', Patrick, R.L.,
        ed., Marcel Dekker, New York, _1_(9), 21, 1967.

Chapter 7

CAN ADHESIVES MEET THE CHALLENGE OF VEHICLE BODYSHELL
CONSTRUCTION ?

I. N. MOODY, P. A. FAY and G. D. SUTHURST

Ford Motor Co. Ltd.,
Research and Engineering Centre, Basildon, Essex

1 INTRODUCTION

Recent advances in adhesive technology such as the ability
to bond directly to oily steel and the emergence of
toughened variants, together with developments in robotic
application methods and new materials have caused motor
manufacturers to reappraise the role of structural adhesives
in future vehicle bodyshell designs.
Current bodyshell design and construction, particularly for
high volume production, is based primarily on the spotwelded
assembly of sheet steel components and highly automated
processes are employed to manufacture large numbers of
bodyshells to repeatably high standards of quality. It is
against this background that adhesives are being evaluated.
It is acknowledged that adhesives could offer advantages of
improved product performance and reduced manufacturing costs
and are prominent contenders for the joining of
multi-material assemblies if their usage is realised in the
future.

2 ADDITIONAL BENEFITS OF ADHESIVES

Provided that adhesives can meet the specific automotive
requirements, then there could be additional benefits such
as:

o   improved strength / stiffness

o   improved NVH (Noise, vibration and harshness) and
      aerodynamics

o   reduced corrosion

o   elimination of spotweld metal finishing operations

o   possible reduction in overall manufacturing costs such
      as combining of joining and sealing operations.

However before any adhesive application can be realised
there are a number of open issues to be resolved :

3   PERFORMANCE REQUIREMENTS

Modern light weight bodyshells make many demands on their
joining techniques. They need to be able to withstand the
normal static loads which hold the components together in
service, in addition to withstanding impact loading
experienced as a result of door slamming, shock loading from
road conditions and the large impact forces generated in
vehicle collisions which may involve high speeds. Under
these conditions the components must hold together whilst
the vehicle kinetic energy is absorbed by yielding and
deformation of the sheet metal.

These static and impact performance requirements need to be
maintained over the broad temperature range experienced on
the road. The body temperature can be as low as $-40^{o}C$ in
winter and up to $+100^{o}C$ under the bonnet after a long run in
hot climates.

Additionally the joints need to be resistant to fatigue
loading generated by road, powertrain and chassis forces and
temperature cycling.

Whilst some basic data (Such as lap shear and T-peel
strengths) can be useful, when available, it is not
sufficient for automotive designers. Joints on a vehicle
structure must last the lifetime of the vehicle but at the
same time be capable of surviving impact conditions for a
fraction of a second, so testing at a few millimetres a
minute does not provide the full picture.

Furthermore, for the automotive industry a comparison is needed between the strength of bonded joints and those produced by existing methods such as spotwelding in which there is a high confidence based on years of field experience. A simple initial comparison can be made from coupon tests. In all the examples that follow adhesives are compared with spotwelds made at 40mm pitch which is an average value across the whole bodyshell. All samples are tested prior to any environmental degradation.

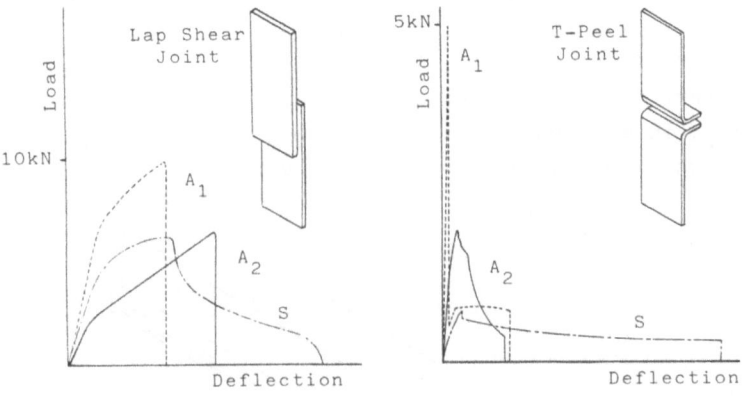

A_1 = Epoxy; A_2 = Acrylic; S = Spotweld

Fig. 1   Results of static lap shear and T-peel coupon tests.

The results of lap shear and T-peel tests on spotwelds and joints bonded using a toughened epoxy (Adhesive 1) and a toughened acrylic (Adhesive 2) are shown in fig. 1.
The spotwelded joint shows a large deflection to failure (ductile failure) due to tearing of the metal around the spotweld to produce a "slug", whereas the adhesives exhibit brittle failures with higher peak loads. The results also show the range of moduli available from different adhesive types.
Simple fatigue results, fig. 2, also show promising results for adhesives with the epoxy giving high fatigue strengths in both lap shear and T-peel, although a wider spread of results.

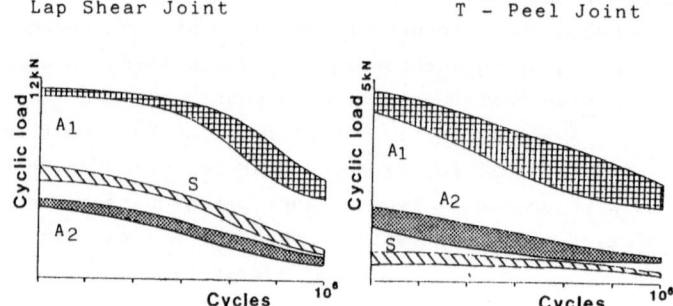

Fig. 2   Results of coupon fatigue tests.

In order to ascertain how these material characteristics
perform in body structures, it is useful to test a
representative component such as a box beam, fig. 3.

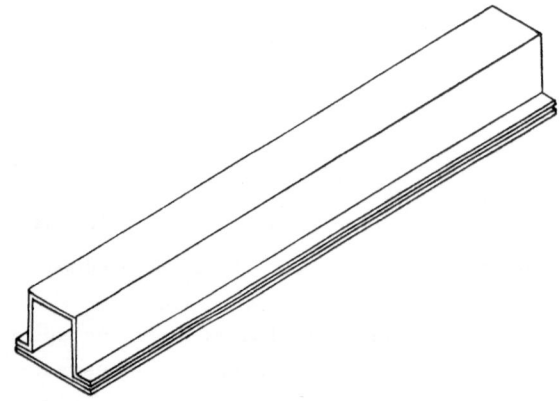

Fig. 3   Schematic of box beam test specimen.

This section is typical of many structural members in a
bodyshell such as crossmembers, siderails, door pillars,
roof rails etc. It can be used for many different types of
tests such as compression, torsion or bending; static,
fatigue or impact; fresh or environmentally degraded joints
and can be tested over a wide temperature range.
Figure 4 shows the results of static box beam bending and
torsion tests showing how the stiffness of a typical section
can be increased using adhesives alone.

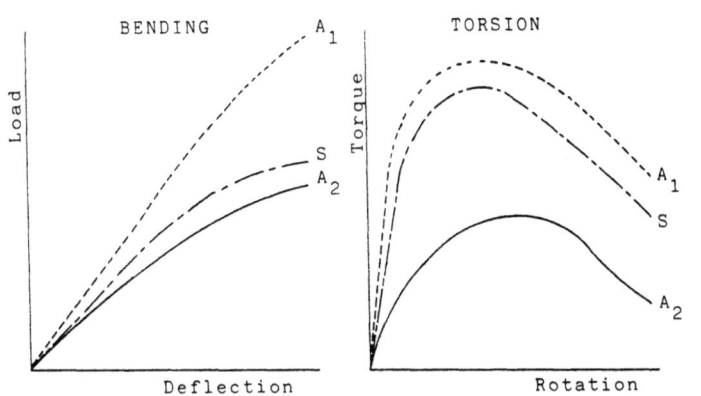

Fig. 4   Results of static box beam bending & torsion tests.

Box beam tests highlight one of the open issues connected
with bonding structural components - impact performance.

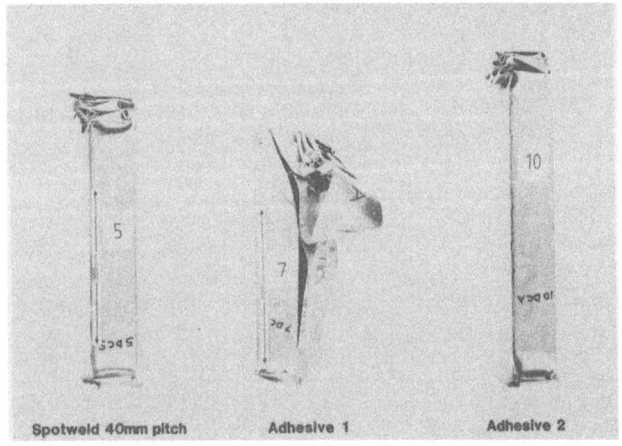

Photograph 1   Box beams after compression impact tests.

Photograph 1 shows the result of axial compression tests
carried out at 30MPH and 20°C. The spotwelded sample
exhibits a stable collapse with regular folds. The tough but
brittle epoxy fails catastrophically (unstable collapse and
'unzipping' along flange length) whilst the more compliant
acrylic gives acceptable collapse.

This concern is increased when the effect of temperature is considered. Photographs 2 & 3 show the variation of impact performance with temperature.

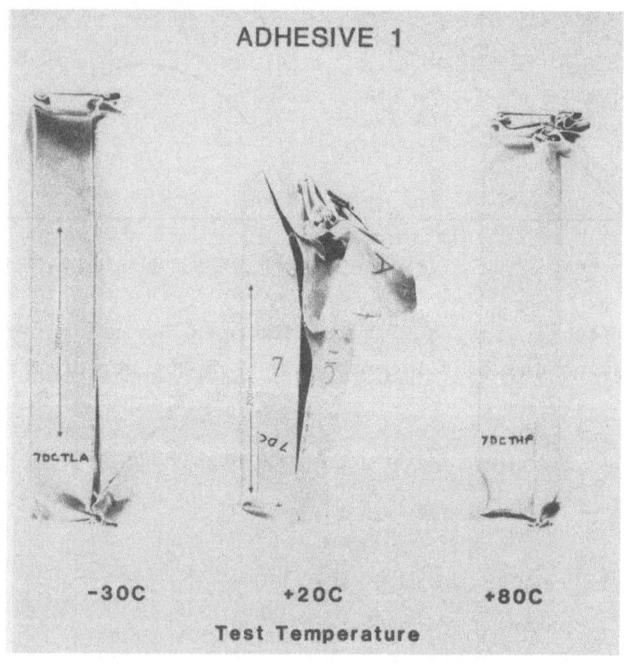

Photograph 2   Effect of test temperature on impact
                performance (Adhesive 1).

At +80°C the epoxy gives acceptable collapse. In comparison the acrylic maintains its acceptable performance at +20 and +80°C, although collapsed lengths vary, but is unacceptable at -30°C.
Similar trends are experienced in impact bending tests, photo. 4 (showing results at +20°C and 20MPH).

Whilst these results show encouraging aspects of adhesive performance they also highlight several shortfalls. The

automotive industry needs a tough adhesive with adequate
impact and fatigue performance which is capable of operating
within the range of service temperatures. Sufficient
technical data on an adhesive should also be available to
enable a comprehensive assessment of its suitability for a
specific application to be made.

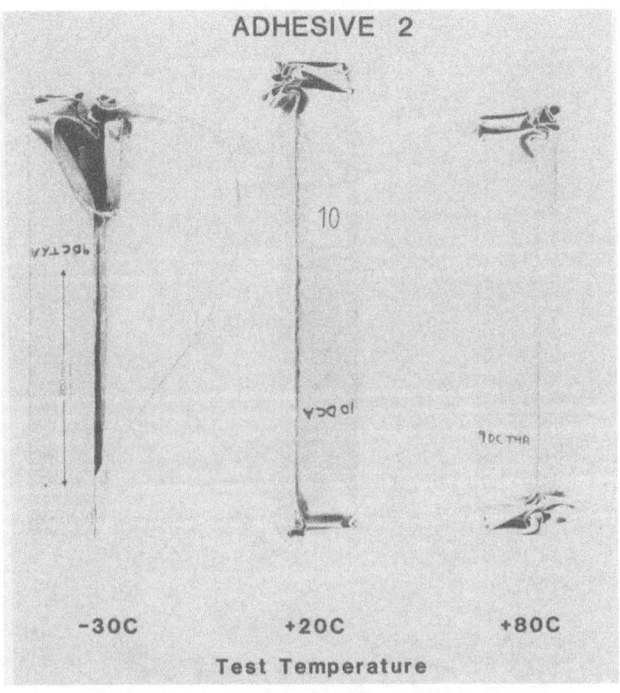

Photograph 3   Effect of test temperature on impact
                performance (Adhesive 2).

4 DURABILITY

Durability is probably the biggest major open issue in the
future use of adhesives on vehicle bodyshells. Adhesives
will not be used in large quantities until a high degree of
confidence in their long term durability is established. The

present uncertainty gives rise to an unacceptable risk
especially in areas which are safety critical or where bond
failure would give rise to catastrophic failure. As
corrosion warranties on new cars are being extended, the
expected service life of bodyshell components is similarly
increased. Thus the demands on bond durability are going to
increase and cannot be compromised.

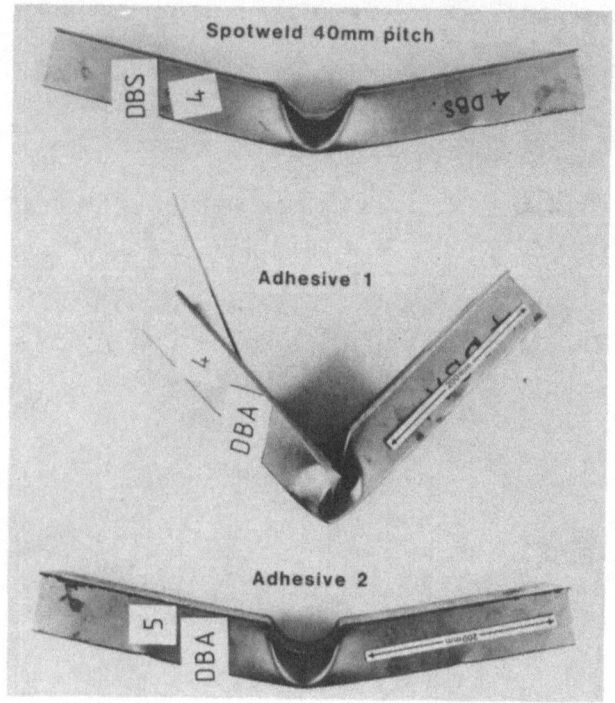

Photograph 4   Box beams after bending impact tests.

The environment in which a vehicle has to operate is varied
and may differ for each vehicle but certain potentially
hostile environments can easily be identified such as water,
salt, petrol, oil, antifreeze, acids, detergents, waxes etc
and the susceptibility of adhesives to these environments

could be determined. Whilst it is easy to be critical of the
results of all accelerated durability tests it is in the
interests of both the adhesive companies and the automotive
industry to jointly develop accelerated test methods which
will help overcome the present confidence gap, otherwise it
will be in the region of 15 years before adhesives will be
used in any highly structural or safety critical areas.
Adhesives manufacturers often claim that these accelerated
tests screen out adhesives with acceptable durability and
are therefore too demanding. The automotive manufacturer's
worry is that an adhesive with unacceptable durability may
manage to pass these tests and fail in service and therefore
tend to specify more demanding tests.
Whilst some durability information may be gained from
building complete vehicles and accumulating large mileages
with them, very little understanding of the mechanisms of
failure and the factors which contribute to or threaten
durability will be established.
Durability of adhesively bonded joints needs to be assessed
in many different ways. The main areas are as follows :
i) Deterioration of properties with time and environmental
degradation
The deterioration of all joint properties during the
expected service life and under service conditions needs to
be quantified. For instance, if the modulus of the adhesives
reduces with water uptake then the stiffness of the
structure will be reduced. Or, as a further illustration,
the impact performance of the joint, and hence the
crashworthiness of the vehicle, needs to be adequate after
many years on the road as well as on freshly made joints.
Other properties that require testing are static strengths
(shear and peel), fatigue strength and other bulk properties
such as cohesive strength of the resin. Any large fall off
in performance is unacceptable and the worst case or lowest
value is essential for design.

ii) Creep strain and rupture

A joint needs to be able to support its service load for the life of the vehicle. A designer needs to know what magnitude of load a joint can be expected to carry for an effectively infinite life, allowing suitable factors of safety. The joint is assumed to have failed if either the bond has ruptured or if the plastic creep strain has caused the component to cease functioning properly through misalignment. Creep becomes more of a concern where the joint experiences constant or near constant loading in high temperature locations, for instance a front suspension mounting, especially if the metal temperature is close to t_g for the adhesive. There appears to be insufficient data available from adhesive suppliers to give confidence in this area.

iii) Visual corrosion

Adhesives must not pose a threat to the current and future levels of corrosion protection and indeed it would be desirable if they could improve on it. The automotive industry has established methods of testing corrosion performance using accelerated corrosion tests involving high humidity, salt and mud together with road driving and stone pecking and these will continue to be useful in assessing the corrosion performance of adhesively bonded vehicles. Simple lab tests have shown that adhesives can be effective at preventing the spread of corrosion through a joint. Corrosion can also be reduced when adhesives eliminate the surface deformation caused by spot or seam welding which may give rise to corrosion, for example, the underside of a roof drip rail.

One of the open issues on corrosion is the feasibility of deleting flange aftersealing operations. It would be desirable if the use of adhesives would provide sufficient protection to the flange edge to remove the need for secondary sealing. Initial tests suggest that this may not always be possible especially where the adhesive fillet is displaced by the wash jets in the paint shop.

Another possible concern exists where a spotweld is made through an adhesive and "sputters" producing voids which lead to subsequent corrosion.

All measures of durability need to take into account the effect of surface treatments and contaminants on performance. Most of the published work on bonding aluminium for bodyshells suggest that pretreatment is necessary (and possible) to achieve strong, durable joints. Less work has been done on oily steel but there seems to be a general concensus that some pretreatment will be required. This needs to be established at a fairly early stage, since it is critical for the introduction of adhesives into mass production where surface preparation is more difficult to achieve. Ideally, if pretreatment is required, it should be applied at the coil stage and should be compatible with press lubricants and mill oils.

5 USE OF ADHESIVES IN MASS PRODUCTION

A modern car assembly plant produces large numbers of quality cars using high levels of automation. Conditions which exist in plant are not necessarily ideal for adhesives and some can be changed, but obviously any large investments in plant and facilities eats into the financial advantages of using adhesives in the first place. Furthermore the need to develop new processes and techniques may involve delay and increase the risk involved with the introduction of new technology.
The two areas of the plant that need consideration are those concerned with the manufacture and finishing of the bodyshell assembly i.e. the Body Stamping and Assembly Shops and the Paint Shop.
The manufacture of the bodyshell is initiated with the delivery of rolls of sheet steel which are subsequently cut into blanks. These in turn are subjected to a series of stamping operations and finally joined together to form the

finished body-in-white. During this process the steel will
have on its surface varying quantities and mixes of mill
oils, drawing compounds, press lubricants and anti-corrosive
coatings.

The predominant joining method in the present process is
spotwelding, carried out either manually, robotically or on
dedicated multiweld fixtures. Manufacturing tolerances exist
on all stamped panels and the final gap between mating
flanges may vary between zero and 3mm.

The spotwelding process has advantages with respect to this
condition as it is capable of 'pulling' the flanges together
thus reducing overall build tolerances.

Visible spotwelds are considered unacceptable in some
instances and a metal finishing operation is performed to
eliminate the surface mark.

In simple terms that is how a body shop operates and this is
the environment in which adhesive application would have to
take place. This imposes many constraints on the adhesives
and corresponding application techniques.

Top of the list is a need for the adhesive to be
toxicologically safe. In an area where there are manual
workers any extra costs incurred in making their use safe
(such as fume extraction, handling precautions, gloves and
masks) have to be taken into the cost equation.

In order to provide the necessary jigging of the bodyshell
and to provide handling strength until the adhesive cures,
it would be desirable to have adhesives which we can
spotweld through. This means that we need to be able to make
adequately strong welds, with no flammability and with no
additional toxicology concerns. These spotwelds have the
additional advantage of acting as peel or creep stoppers in
critical areas.

Having already mentioned the large gaps which may occur in
assembly, then an adhesive needs to be gap filling because
the bonding operation does not 'pull' flanges together in
the way that spotwelding does. The implications of this

requirement are that there will be little or no drop off in properties with increasing adhesive thickness and that the adhesive material will not "slump" or "sag" out of the joint before the adhesive is cured. If the adhesive requires oven curing in the paint shop, then the time delay between application and cure will certainly be hours, and may even be days. There is also a possibility of sag on the temperature climb up when the bodyshell enters the oven. Robotic application is considered essential in the future use of adhesives. The robotic system will be required to accurately measure quantities of adhesive dispensed, perform any necessary mixing, apply the adhesive to the panel and do this sufficiently quickly to maintain production speeds, allowing for changes in the ambient plant temperature. The shelf life of the adhesive under normal plant operating conditions should typically be about 6 months. Shorter shelf lives may be tolérable but any essential storing conditions need to be carefully explained.

Inevitably, some adhesive material will be squeezed out of the joints when the two panels are brought together. Adhesive squeeze-out has a big effect on the visual appearance of the joint and may make the fitting of subsequent parts more difficult. Clearly, methods must be developed for avoiding or dealing with this concern. These methods may be simple design changes, fig. 5, so that adhesive is spewed out into non-visible areas or adhesive developments whereby the adhesive can quickly and easily be removed, possibly after partial cure or "gelling".

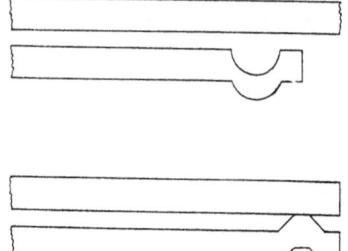

Fig. 5   Design changes to reduce squeeze-out.

It is also essential to understand the level of
compatability between an adhesive and the numerous types of
mill oils, etc that could accumulate during the stamping
process. Whilst it is possible to reduce the number of
different materials, it would be easier if the adhesive was
not too sensitive to different ones. It is important to keep
an eye on future developments in this field such as dry film
lubrication.

Once the body assembly is completed it then progresses to
the paint shop. The current process is outlined in fig. 6.

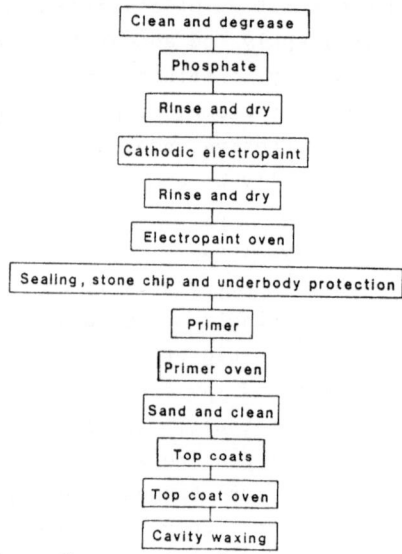

Fig. 6   Overview of current paint process.

The first stage in the paint shop is a thorough clean and
degrease which involves chemical cleaning solutions and high
power water jets. This thorough cleaning is essential for
the corrosion protection of the vehicle and is unlikely to
be reduced in ferocity. Unfortunately, the jets are powerful
enough to displace uncured adhesive. This is a highly
undesirable situation because, in addition to removing
adhesive from the joint, it could end up on the middle of a

skin panel which would be difficult to clean off. Disturbing
the adhesive fillet can also reduce the corrosion protection
offered to an exposed flange edge as mentioned earlier.
There are many potential solutions to this problem. One
method is to increase the wash resistance of the adhesive by
formulation changes. Or the adhesive could be partially or
fully cured before reaching the washes. It could be cured by
induction or full oven cure (but beware of the effect of
baking oils on to the metal surface!) or be a two part
adhesive with room temperature cure. Film or tape adhesives
may also solve this concern. All these solutions may also
help the sag resistance concerns described earlier. Design
changes might help the adhesive stay out of direct
impingement of the jets in some areas, but cannot be relied
upon as a universal solution.

Following cleaning, the body is then coated with zinc
phosphate before being treated with cathodic electropaint.
In this process the complete bodyshell is negatively charged
and totally submerged in a bath of positively charged paint.
The coated body is rinsed and blown dry before entering the
stoving oven.

A typical electrocoat cycle runs at 175°C for ten minutes,
although some caution needs to be exercised in quoting
stoving times and temperatures. The bodyshell acts as a
large heat sink and not all areas of the body warm up at the
same rate. So, although all areas see the minimum cycle,
some may see longer and/or hotter cycles. The situation is
further confused because the bodies are on a moving
production line which, for a number of reasons, does not
always move continuously. When this happens the heat input
to the oven is turned off, but the metal temperature may
rise at first and maximum temperatures of about 200°C are
possible. The other effect of line stoppages is that certain
cars may see longer cycles still.

Following electrocoat, various sealing operations are
carried out before the application of primer. The primer is
baked for a minimum cycle of 165°C for 5 minutes, with all
the possible variations described above.

Next, the body is painted with the top coats and baked at a lower temperature - $120^{\circ}C$ for 10 minutes.

The body receives the final operations such as cavity filling before being trimmed and assembled.

It is essential that the adhesive is completely compatible with the paint system. There are many aspects to this :

1) The adhesive must be fully cured on leaving the paint shop.

2) The adhesive must not degrade if it sees overbake conditions such as a body which needs repair to primer or enamel coats and sees that part of the cycle twice.

3) The finished painted joint must be visually acceptable. The adhesive must not stain the paint and the paint must take adequately to it.

4) The adhesive performance must not be reduced as a result of the painting operations.

5) The adhesive must not dissolve in electrocoat and so contaminate the tank.

6) It would be desirable if the adhesive was sufficiently conductive to take electrocoat. Sufficient spotwelds or earthing straps need to used to ensure that no panels are electrically isolated.

7) The adhesive needs to meet all the weatherability tests for the paint system.

To use adhesives in mass production it is necessary to have good in-house quality control systems in addition to those of the suppliers. It will be necessary to ensure that the adhesive is within specification, properly mixed, applied to right parts in the right amounts and cured satisfactorily. Also required is a non-destructive test method which can determine the presence of a satisfactory bond, rather than just checking on the presence of adhesive.

Whilst this manufacturing process is not ideal for adhesives, if the benefits are worthwhile, then parts of the process, design and adhesive formulation can be altered to realise the benefits already described. Once again, though, changes in process mean capital investment and they will not take place without justification.

6 COST EFFECTIVENESS OF ADHESIVES

The financial implications of using adhesives are complex.
The final calculation needs to take into account such
factors as :-

o   Material costs
o   Manufacturing costs such as:
    o   Elimination of spotwelding and metal
         dressing operations
    o   Elimination of sealing operations
    o   Reduction in assembly times
    o   Additional investment required : robotic
         application equipment, storage facilities,
         curing equipment, metal pretreatment
         equipment, tighter process control and NDT
    o   Manpower costs
    o   Maintainance costs
o   Implications on product design and performance.

Whilst this list is not exhaustive, it shows how complex the
analysis is. Adhesives will need to show significant
benefits in either manufacturing costs or product
performance before the mass production motor industry will
consider changing over to bonding.

7 DESIGN IMPLICATIONS

Finally it is necessary to consider the implications which
adhesives have on the design process.
It can easily be shown that adhesives have the potential to
increase freedom in design. For instance, in order to make a
spotweld access is needed to both sides of a joint and
welding flanges need to be external to closed sections.
Possible alternative section designs which use adhesives to
overcome these limitations are shown in fig. 7.
Looking to the future, adhesives can widen the mix of

materials available to the body designer to form a bodyshell
assembly such as composites, aluminium, coated steels,
plastics. Using mixed material assembly, such as hang on
panels on a metal frame, may allow restyling at much lower
costs and adhesives could obviously play a major role in
this type of development.

SPOTWELDED   **ADHERED**

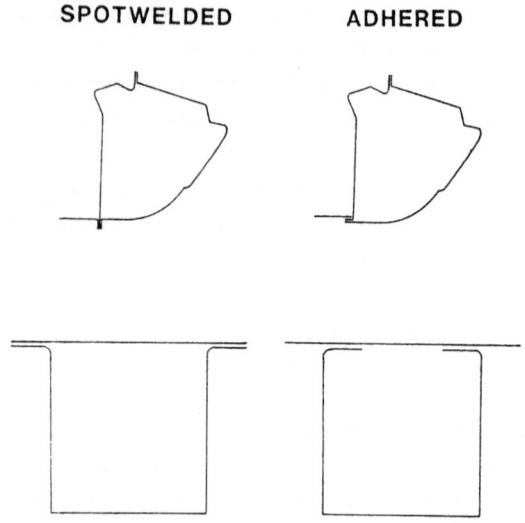

Fig. 7   Alternative section designs using adhesives.

The higher stiffnesses available on bonded structures may be
used to reduce vehicle weight (and hence fuel consumption).
Furthermore it is possible to tailor the adhesive properties
for different requirements on different parts of the
bodyshell. A spotweld is always a spotweld but different
types of adhesives vary from the very stiff to the highly
compliant. This variation of properties could be used
effectively to increase the overall performance of the
bodyshell.
Having described some of the potential design advantages it
is important to know how these advantages can be realised.
The design of spotwelded sheet steel body structures has

evolved over a considerable number of years and a wealth of experience and knowledge is now available. This is used today in conjunction with the latest technology in computers and finite element analysis in the development of new body designs.

Adhesives need to 'catch up' and a considerable amount of background work needs to be conducted to gain a similar working knowledge for adhesives.

The performance of joints and sections under typical complex loading modes needs to be established. A lot of the published work in this area describes joints with simple uni-axial loading or describes compressive loading of adhesives to such a degree that the adhesive is not required at all!

The current efforts seem to be misdirected with the motor companies testing adhesives to establish performance criteria and the adhesive companies suggesting vehicle redesign when a performance shortfall is identified.

A more beneficial use of industrial expertise would result if these roles were exchanged.

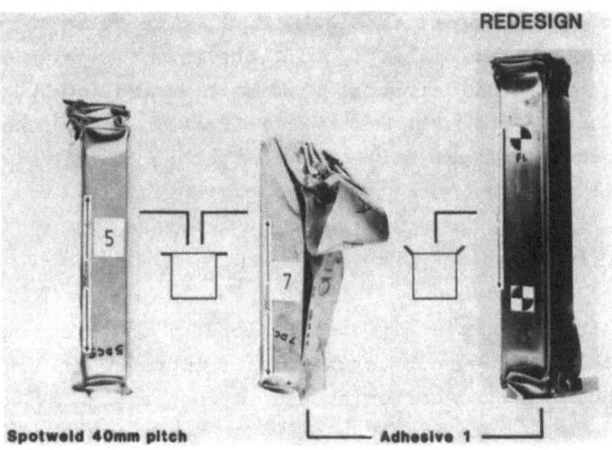

Photograph 5   Redesigned box beam for axial collapse.

Having said that, quite simple changes in design can be made to dramatically improve the performance of a bonded

structure. Returning to the box beam impact test a simple
redesign, in this case turning the flanges through 45°,
photo. 5, can make the performance of even a brittle epoxy
acceptable. It must be remembered however that a redesigned
structure will inhibit parallel development of designs
compatible with either spotwelding or adhesives.

All design is a compromise of requirements and so it is not
always possible to produce the "ideal" design for adhesives.
Other requirements such as weight, sectional properties,
complexity of pressings, ease of assembly and cost must also
be considered. Whilst the optimum design for adhesives may
not be acceptable for other reasons, an improved design to
maximise the benefits may well be achievable.

There is a need for adhesive bulk data for use in FEA and
other design aids. Simple properties such as yield, modulus
and stress/strain to failure are essential engineering
requirements and as such need to be established. A lot of
controversy seems to exist over test methods, but the type
of test method may be considered largely academic provided
that it produces accurate, repeatable results which can be
validated by component testing. More work needs to be
carried out in the area of joint analysis, particularly
involving the non-linearities of both adhesive and substrate.
Finally, on the design side, if adhesives are introduced in
major structural areas then large factors of safety will be
used until greater confidence is gained.

8   CONCLUSIONS

This is not a completely dismal picture; there are many
encouraging signs such as the impact improvement due to
redesign, improved corrosion due to the continuous sealing
effect of the adhesive, the increased stiffness and
strength, the improved fatigue strengths and the elimination
of visible spotwelds and we believe that many of the
remaining open issues can be resolved.

It will require however, a combined effort from the adhesive

industry and the motor industry to develop an acceptable
joining process which will almost certainly involve
modifications to both the body design and the manufacturing
process as well as the development of improved adhesives.
Future body construction trends will inevitably include an
increasing percentage of different materials which will
necessitate the introduction of different joining techniques
and obviously adhesives must figure prominently as
contenders for this role.

In conclusion, it's true to say that we have the requirement
for an adhesive but what remains in doubt is 'Do we have the
necessary adhesive technology ?'

Until the open issues that we have outlined have been
resolved and we have confidence in the performance of
adhesives in relation to our product then the completely
adhered body shell will remain very much a concept rather
than a reality.

Chapter 8

PROPERTIES OF ADHESIVES FOR COMPOSITE AND BONDED METAL REPAIRS

K.B.ARMSTRONG

Britsh Airways
P.O.Box 10, London Airport, Hounslow, Middx, TW6 2JA

INTRODUCTION

The choice of adhesives for the repair of composites and bonded metal parts needs careful consideration. Ideally a repair should be carried out using the same adhesive as that used for the original construction. A close match of properties is important because harder and more brittle resins have lower impact resistance and give lower joint strengths than tougher resins. The elongation of the material under repair needs to be matched by the elongation of the resin. A high modulus material can use a higher modulus adhesive than a low modulus material. Ref.1. However, original manufacture commonly involves hot-curing film adhesives or matrices. For repairs this raises problems of refrigerated storage of adhesives or pre-pregs.

In an Airline or Air Force several types of aircraft may be used. If composites or bonded metal parts from several different manufacturers have been purchased the question arises, whether all can be repaired with the same materials? This problem is further compounded by the fact that fabrics of different materials, weaves and weights may have been employed. This has led some users to consider stocking one film adhesive for each cure temperature and to interleave these between layers of dry fabric of the type required for each repair. It has the advantage that the same film adhesive can then be used for composite and bonded metal repairs, avoiding the need to keep a wide range of materials under refrigerated storage. It does, however, raise the question of how the properties of the chosen film adhesive compare with the properties of the resin systems used in the various composites.

Unfortunately manufacturers commonly do not quote all of the relevant properties either for composite matrix resins or film adhesives. The provision of more data is to be most warmly encouraged.

Comprehensive data would usefully include for fully cured material :-

1. Tensile strength
2. Compression strength
3. Tensile modulus
4. Compression modulus
5. Shear strength
6. Shear modulus
7. Elongation at failure
8. Water Diffusion Coefficient

9. Water Solubility Coefficient
10. Creep properties
11. Fatigue properties

Ideally all this data should cover the whole of the service temperature range.

Together with :-

12. Glass transition temperature after cure at various temperatures
13. Glass transition temperature at various water uptake levels
14. Coefficient of thermal expansion

Items 1 - 7 are very dependent on testing technique and it is important to use standard methods and specimens. One variable, very difficult to cover, is the range of strain rates in service. For aircraft these can vary from very slow creep rates, possibly while parked in hot sun, to very high rates due to gust loading in turbulent air perhaps occurring at low temperatures. Hence it is important to control and specify temperatures at which testing is carried out.

The modulus values in particular will affect the behaviour of the composite, as a whole, in bending or compression buckling. Strength is considerably reduced above the glass transition temperature of the resin and T_g can be reduced by as much as 20°C for each 1% of water absorbed. Fig.1, Ref.10 (Delmonte), Ref.13, Wright. These problems are aggravated if cold-setting adhesives are used because their T_g is seldom much above their cure temperature and their water uptake can be greater than most hot-sets. T_g can usually be increased by warm curing, say at 50°C or 80°C. Table 1 (Courtesy of RAE).

In practical situations it is often difficult, to carry out a repair at the original cure temperature. To do so would mean extensive tooling. It is also necessary to carry out the curing at a fairly precise and controlled temperature. Overheating one area to achieve the minimum temperature in another may make the hotter area too brittle.

If pressure above vacuum cannot be applied in a practical case and/or tooling is not available, a technique, beginning to find favour with aircraft manufacturers is to use an adhesive that can be cured at a temperature lower than that at which the part was originally made. For example a part made at 180°C can be repaired using a 120°C curing adhesive.

The work reported in this paper was primarily directed at the repair of composites with cold-setting adhesives and also provides guidance on the choice of adhesives for bonded metal repairs. The objective was to obtain tensile strength, tensile modulus, and elongation to failure and the various figures indicate that these fundamental resin properties can be related to lap joint performance and hence used for the development of better adhesives. A further objective was to obtain Diffusion and Solubility Coefficients in order to assess the likely reduction of T_g.

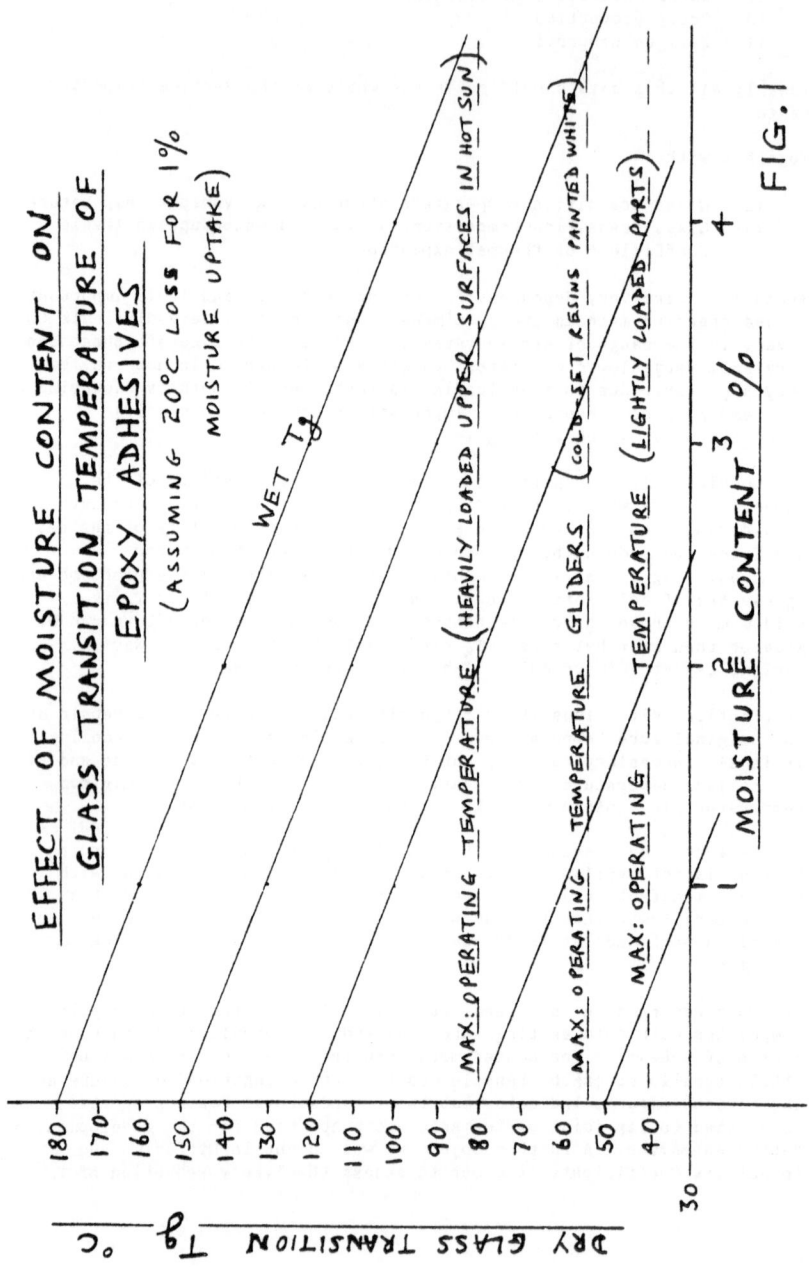

FIG. I

TABLE 1

ADHESIVE	T_g DRY & WET AFTER VARIOUS CURE TEMPERATURES					
	After RT Cure Dry	After RT Cure Wet	After 50°C Dry	After 50°C Wet	After 80°C Dry	After 80°C Wet
EA 9330	11°C				27°C	
EPIKOTE 815 + RTU	48°C		61°C		69°C	
EPIKOTE 828 + VERSAMID 125	47°C		51°C			47°C
EC 2216	**				**	
EC 3524	**				**	
EA 9321					**	
AF 163 *	---	---	---	---	108°C	82°C
EC 3559					71°C	
EA 9330 + 20% MICROBALLOONS						
EC 3568			---	---	---	---
EC 3578					**	
PERMABOND E34					**	

* AF 163-2M film adhesive cured at 120°C for 1 hour
The above results were obtained using D.S.C. Where results were not obtainable by this method, indicated by ** it is intended to try the torsion pendulum method. Other blank spaces indicate that work has not yet been carried out.

It was hoped to show that, provided the T_g values remained acceptable, to meet the service temperature requirements, then some cold-setting adhesives might have adequate mechanical properties for repair work and thus make these tasks a great deal easier.
Since this work was started some new two-part adhesives have begun to come on the market for composite repairs. They can offer a T_g of about 120°C from a room temperature cure. If their equilibrium water uptake can be limited to 2% or less then they should be quite suitable for permanent repairs to subsonic aircraft and also to supersonic aircraft if the skin temperature does not exceed 80°C. This assumes that they can also provide the required mechanical properties.

Cold-setting repairs can often be done with very little tooling and require no more than vacuum pressure. Also radiant lamp or heater mat curing is sufficient to improve both T_g values and the speed of the repair.The repair of contaminated composites may also need to be considered.See Ref.33.

TEST PROGRAMME

Initially, four cold-setting adhesives and one potting compound were tested. The materials selected were:

    Hysol EA 9330
    Shell Epikote 815 + Epikure RTU
    Shell Epikote 828 + Versamid 125
    3M EC2216
    3M EC3524.

Specimens were of the shape and dimensions shown in Fig.2. Data reported was related to the 100mm gauge length. Elongation to failure was taken from crosshead motion measured by the chart recorder. In most cases bubbles were present and failure often occurred through these so it might be more accurate, to take the highest value of the three specimens in each case as the most nearly correct for a thin film without bubbles.

Subsequently other adhesives were added to the programme. They were:

    Hysol EA 9321
    3M AF163-2M
    3M EC3559
    Hysol EA 9330 + 20% microballoons
    3M EC3568
    3M EC3578
    Permabond E34

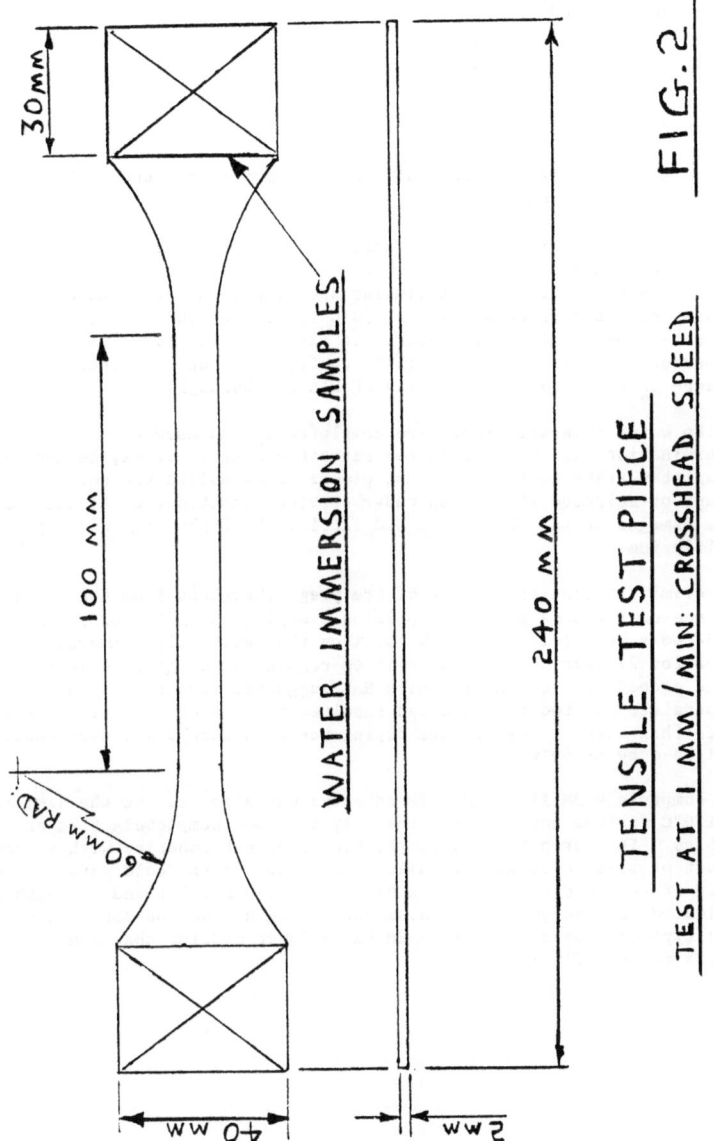

WATER IMMERSION SAMPLES

TENSILE TEST PIECE

TEST AT 1 MM / MIN: CROSSHEAD SPEED

FIG. 2

Tensile tests were carried out to A.S.T.M. D–638 on each specimen and from these the tensile strength and elongation at failure were obtained from the chart record. Young's Modulus (E) was obtained from the slope of the chart record and the Shear Modulus (G) was calculated from this using the standard formula:

$$G = \frac{E}{2(1 + \nu)}$$

where ν = Poisson's Ratio

Strain Energy at failure was calculated from the area under the stress/strain curve.

The data obtained is listed in Tables 2 - 5. The specimens were put to further use in that 30mm lengths were cut from each end of the test pieces. For each set of six similar specimens, one pair was examined as it was, one pair after heating at 50°C for three hours and the third after heating at 80°C for 1 hour. In each case the first piece was used to measure Diffusion and Solubility Coefficients and the second to measure T_g and hardness (Courtesy of RAE Farnborough).

As the water immersion tests are completed T_g and hardness determination will be made in the saturated state. It may be necessary, at a later stage to fly some test pieces to establish the equilibrium amount of moisture absorption under service conditions or to cut samples from damaged parts, which have had sufficient flight time and dry them to equilibrium.

A preliminary test on a piece of pre-preg fibreglass from the inside surface of a radome proved surprising. Drying at 150°C for several days produced a weight loss of 1.7% (5.1% on the resin only).Immersion in water for 21 days did not cause it to regain its original weight let alone exceed it. Discussion with RAE suggested that the drying temperature was too high and had resulted in loss of volatile material other than water. Any further drying tests to assess moisture content will be done at 60°C.

For comparison 3M–AF163-2M film adhesive was also cast in the same mould at 120°C for one hour and was the only specimen completely free of bubbles. It proved to be a tough, high strength adhesive with a greater amount of plastic strain at failure than most of the cold–sets. However, a few of the cold-sets showed a higher tensile modulus and strength and indicated that adequate stiffness and strength could be obtained from this type of product. 3M–AF163-2M has a lower modulus than composite matrix resins (Fig.11A).

TABLE 2

SPECIMEN NO	ADHESIVE	ADHESIVE TENSILE STRENGTH		REMARKS
		PSI	MPa	
1	HYSOL EA 9330	1,377	9.5	Many small bubbles. Cross head speed 0.45mm/min Material soft - mix suspect
2	HYSOL EA 9330	3,659	25.25	Bubbles upper face only
3	HYSOL EA 9330	4,420	30.5	Bubbles upper face only
4	EPIKOTE 815 + RTU	7,174	49.5	A few bubbles
5	EPIKOTE 815 + RTU	10,145	70	A few bubbles
6	EPIKOTE 815 + RTU	10,145	70	A few bubbles
7	EPIKOTE 820 + VERSAMID 125	7,246	50	A few large bubbles mainly at one end
8	EPIKOTE 828 + VERSAMID 125	6,812	47	No bubbles in gauge length
9	EPIKOTE 828 + VERSAMID 125	6,666	46	One bubble at the edge, about centre. Failed across this bubble

TABLE 2

SPECIMEN NO	ADHESIVE	ADHESIVE TENSILE STRENGTH		REMARKS
		PSI	MPa	
10	3M EC2216	1,480	10.25	Failed across large central bubble
11	3M EC2216	1,880	13	Failed from edge crack due to bubble hole
12	3M EC2216	1,558	10.75	Failed from edge crack
13	3M EC3524	1,377	9.5	Brittle failure
14	3M EC3524	1,377	9.5	Brittle failure
15	3M EC3524	1,232	8.5	Damaged edge. Failed away from damage
16	Hysol EA9321	3,985	27.5	Small bubbles
17	Hysol EA9321	6,087	42	No bubbles
18	Hysol EA9321	4,348	30	1mm dia bubbles

TABLE 2

SPECIMEN NO	ADHESIVE	ADHESIVE TENSILE STRENGTH		REMARKS
		PSI	MPa	
19	3M AF163	6,340	43.75	120°C cure, film adhesive. No bubbles
25	3M-EC3559	3,405	23.5	
26	3M-EC3559	3,478	24	
27	3M-EC3559	4,058	28	
28	HYSOL EA9330 + 20% MICRO-BALLOONS	725	5	.01 strain rate
29	HYSOL EA9330 + 20% MICRO-BALLOONS	1,739	12	.1 strain rate
30	HYSOL EA9330 + 20% MICRO BALLOONS	725	5	.02 strain rate
31	3M-EC3568	2,391	16.5	.01 strain rate
32	3M-EC3578	1,920	13.25	Failed at edge bubble
33	3M-EC3578	1,522	10.5	Failed from internal bubble at edge
34	3M-EC3578	2,536	17.5	
35	PERMABOND E34	3,913	27.0	
36	PERMABOND E34	2,971	20.5	Large bubble at failure point

TABLE 3

ADHESIVE	LAP SHEAR PSI	E PSI	$\sqrt{G}$	G CALCULATED PSI	ADHESIVE TENSILE STRENGTH $\div \sqrt{G}$
EA9330	3,500	84,000	173	30,000	23
EA9321	4,190	125,000	217	47,000	20.7
EPIKOTE 828 VERSAMID 125	2,180	112,000	5	42,000	33.7
EPIKOTE 815 + RTU	783	178,000	259	67,000	38.6
EC2216	1,920	3,100	32	1,030	50
EC3559	5,450	94,000	181	33,000	19.9
AF163	5,000	95,650	184	34,000	34.5
EC3524	1,350	33,000	105	11,000	12.4
EC3568	3,150	98,550	187	35,000	12.79
EC3578	2,878	87,000 av	175	30,500	14.5
PERMABOND E34	2,174	182,500	262	68,500	14.9

TABLE 4A

ADHESIVE	After RT Cure "D"	After RT Cure "S"	After 50°C Cure "D"	After 50°C Cure "S"	After 80°C Cure "D"	After 80°C Cure "S"
	DIFFUSION COEFFICIENT "D" $m^2.s^{-1}$ SOLUBILITY COEFFICIENT "S"					
EA 9330	1.63^{-13}	13%	1.59^{-13}	17.2%	1.59^{-13}	10%
EPIKOTE 815 + RTU	3.88^{-14}	7.35%	4.62^{-14}	4.5%	5.50^{-14}	3.2%
EPIKOTE 828 + VERSAMID 125	3.88^{-14}	6.25%	5.10^{-14}	6.1%		
EC 2216	1.63^{-13}	5.7%	1.19^{-13}	5.7%	1.45^{-13}	4.7%
EC 3524	1.3^{-13}	35%	1.19^{-13}	32.0%	1.13^{-13}	33.6%
EA 9321	9.34^{-14}	7%	4.33^{-14}	4.35%		
AF 163 *	- - -	- -...	- -	- - ..	9.34^{-14}	1.4%
EC 3559	9.34^{-14}	4.7%	6.49^{-14}	4.30%		
EA 9330 + 20% MICROBALLOONS	7.54^{-14}	81%	4.52^{-14}	56.6%	9.10^{-14}	104%
EC 3568	**	16.77%	---	- - -	--.-	- -..
EC 3578	**	8.10%	**	14.6%	**	7.35%
PERMABOND E34	**	1.50%	**	1.53%	**	1.72%

* AF 163-2M film adhesive cured at 120°C for 1 hour
** These resins gave a sigmoidal curve of M_t/M_∞ against the square root of time, indicating non-Fickian diffusion. D could not be obtained from these curves.
NOTE! The 17.2% uptake for EA 9330, post-cured at 50°C, is suspect. This specimen was softer than the other two when tested after room temperature cure before post-curing for the water immersion test.

ADHESIVE & TEMPERATURE	"D" m^2s^{-1}	"S" %	REMARKS
CIBA GEIGY			Hot setting film
BSL 312　　　50°C	3.0×10^{-13}	2.2	120°C
25°C	1.4×10^{-13}	2.1	
1°C	1.3×10^{-14}	2.6	
AMERICAN CYANAMID			Hot setting film
FM1000　　　50°C	3.2×10^{-12}	15	120°C
25°C	1.1×10^{-12}	16	
1°C	7.5×10^{-14}	21	
AMERICAN CYANAMID			Hot setting film
FM73　　　60°C	1.37×10^{-12}	2.7	120°C
100°C		12.0	
CIBA GEIGY			Farnborough result .6mm
BSL 312　　　50°C	12.5×10^{-13}	2.8	thick sample
CIBA GEIGY			Two-part R.T. cure 1.9mm
Redux 410　　50°C	16×10^{-13}	4.4	thick Farnborough result
CIBA GEIGY			1.3mm thick Farnborough
Redux 410　　50°C	10×10^{-13}	5.2	result
CIBA GEIGY			.3mm thick Farnborough
Redux 410　　50°C	7×10^{-13}	3.6	result
HYSOL			Two-part R.T. cure .8mm
EA 9309.2　　50°C	6.2×10^{-13}	4.1	thick Farnborough result
ARALDITE MY790 + HY951			Cured 3 hours at 60°C
84°C	1.0×10^{-12}	5	"　　)
60°C	7.0×10^{-13}	3.9	"　　)
40°C	4.4×10^{-13}	3.1	"　　) Two Part Mix
21°C	1.0×10^{-13}	2.4	"　　)
3°C	5.0×10^{-14}	1.8	"　　)
			From p190 Int.Journal
			of Adhesion &
			Adhesives Oct 1983

131

MOISTURE ABSORPTION PROPERTIES OF COMPOSITE MATRIX RESINS

(FROM REF.13)

ADHESIVE & CURING AGENT	MANUFACTURER	TEMP °C	ABSORPTION D $m^2.s^{-1} \times 10^{14}$	S%	RH %
EPIKOTE 828 +TETA		23	–	3.92	Immersion
		45	–	3.90	Immersion
		75	–	4.12	Immersion
MY 750 + DDM	CIBA GEIGY	0.2	2.67	2.59	100
		25	20.9	2.51	100
		37	40.9	2.36	100
		50	102	2.47	100
		60	179	2.44	100
		70	316	2.45	100
		80	411	2.57	100
		90	630	2.77	100
MY 750 + PA	CIBA GEIGY	37	13.2	1.35	100
5208	NARMCO	30	–	5.93	100
		75	–	6.10	100
		100	–	5.98	100

132

TABLE 4C

ADHESIVE & CURING AGENT	MANUFACTURER	TEMP °C	ABSORPTION D $m^2.s^{-1} \times 10^{14}$	S%	RH %
EPON 1031/BDMA	SHELL	30	–	1.67	80
DER 332 + DDS		30	–	2.84	80
		45	–	2.61	80
		60	–	2.48	80
		75	–	2.07	80
X-904 + ANHYDRIDE	FIBERITE	25	113	1.1	70
X-915	FIBERITE	25	–	3.8	70
X-934 + AMINE	FIBERITE	25	–	4.8	75
E-293 + ANHYDRIDE	FERRO	25	47.7	0.9	75
E350 + AROMATIC AMINE	FERRO	25	7.44	3.1	75
E450 + AROMATIC AMINE	FERRO	25	–	4.6	75
X-2003 + AMINE	HERCULES	25	–	1.6	75
3002 + AMINE	HERCULES	25	3.53	7.7	75
SR-10500	WHITTAKER	25	–	2.3	75
1004	WHITTAKER	25	–	2.8	75
MY720 DDS/BF$_3$-MEA	CIBA GEIGY	70	–	5.6	100

ADHESIVE & CURING AGENT	MANUFACTURER	TEMP °C	ABSORPTION D $m^2.s^{-1} \times 10^{14}$	S%	RH %
3501-5 + AMINE	HERCULES	NOT GIVEN		6.2	100
3501-6 + AMINE	HERCULES	NOT GIVEN		6.2	100
3502	HERCULES	NOT GIVEN		6.3	100
934	FIBERITE	NOT GIVEN		6.5	100
NMD 2373		NOT GIVEN		9.9	100
5208	NARMCO	NOT GIVEN		6.3	100
5208	NARMCO	71		7.0	IMMERSION
3501-5 + AMINE	HERCULES	49		5.9	IMMERSION
		71		5.9	IMMERSION
5208	NARMCO	49		6.4	IMMERSION
5209	NARMCO	25		3.17	99
		51		4.53	92
		64		4.55	97
DER 332 + DDS		RT		4.14	95
934	FIBERITE	RT		4.69	95

134

PAGE 4 of 4

ADHESIVE & CURING AGENT	MANUFACTURER	TEMP °C	ABSORPTION D $m^2.s^{-1} \times 10^{14}$	S%	RH %
X80 + MNA		RT		3.48	95
XD7342 + DDS		RT		5.4	95
5208	NARMCO	RT		6.4	95
X801 + TONOX		RT		7.31	95
X801 + DDS		RT		7.15	95
904	FIBERITE	RT		1.71	95
EPON 1031 + MNA	SHELL	RT		2.98	95

135

TABLE 4D   PAGE 1 of 2

DIFFUSION COEFFICIENTS FOR VARIOUS COMPOSITE MATRIX RESINS
AT 100% RH OR IMMERSION (FROM REF.13)

ADHESIVE & CURING AGENT	MANUFACTURER	TEMP °C	D $m^2.s^{-1} \times 10^{14}$
MY750 + DDM	CIBA GEIGY	0.2	2.67
		25	20.9
		37	40.9
		50	102
MY750 + PA	CIBA GEIGY	37	13.2
3501-5 + DDS	HERCULES	23	7.65
		49	31.0
3501-6 + DDS	HERCULES	23	4.59
		60	36.1
3502	HERCULES	23	4.96
		60	37.3

DIFFUSION COEFFICIENTS FOR VARIOUS COMPOSITE MATRIX RESINS
AT 100% RH OR IMMERSION (FROM REF.13)

ADHESIVE & CURING AGENT	MANUFACTURER	TEMP °C	D $m^2 s^{-1} \times 10^{14}$
5208 + DDS	NARMCO	23	3.61
		60	36.9
934 + DDS	FIBERITE	23	3.01
		60	20.8
NMD 2373		23	2.18
NMD 2373		60	36.3
3501-5 + DDS	HERCULES	49	31.8
5208 + DDS	NARMCO	49	43.3
		30	15.0

Where several results are quoted for the same resin they came
from different sources.

137

TABLE 4E

MOISTURE ABSORPTION PROPERTIES OF SOME HOT-CURED FILM ADHESIVES

ADHESIVE	MANUFACTURER	TEST TEMP °C	ABSORPTION "D" $m^2.s^{-1} \times 10^{14}$	"S"	RH %
BSL 312	CIBA GEIGY	50	30	2.2	IMMERSION
		25	14	2.1	IMMERSION
		1	1.3	2.1	IMMERSION
FM 1000	AMERICAN CYANAMID	50	320	15	IMMERSION
		25	110	16	IMMERSION
		1	7.5	21	IMMERSION
FM 73	AMERICAN CYANAMID	100	-	12	100
		65	137	2.7	100

TABLE 4F

MOISTURE ABSORPTION PROPERTIES OF SOME TWO-PART COLD-SETTING ADHESIVES

ADHESIVE	MANUFACTURER	TEST TEMP °C	ABSORPTION "D" $m^2.s^{-1} \times 10^{14}$	"S"	RH %
REDUX 410	CIBA GEIGY	50	160	4.4	96
EA 9309.2	HYSOL	50	62.0	4.1	96
EA 9330	HYSOL	RT	16.32	13	IMMERSION
EPIKOTE 815 +RTU	SHELL	RT	3.88	7.35	IMMERSION
EPIKOTE 828 + VERSAMID 125	SHELL GENERAL MILLS	RT	3.88	7.35	IMMERSION
EC 2216	3M	RT	16.32	5.7	IMMERSION
EC 3524	3M	RT	13	35	IMMERSION

TABLE 5
COMPARISON OF ADHESIVES BY VARIOUS MECHANICAL & PHYSICAL PROPERTIES - ROOM TEMPERATURE DATA

ADHESIVE	LAP SHEAR STRENGTH ON ANODISED AL.ALLOY	TENSILE STRENGTH OF RESIN	TENSILE MODULUS OF RESIN	STRAIN TO FAILURE TOTAL	PLASTIC STRAIN AT FAILURE	WATER SOLUBILITY COEFFICIENT	STRAIN ENERGY AT FAILURE N.mm
EA9330	3,500	4,000	84,000	6.5%	1% (2) 3% (3)	13%	20.435
EA9321	4,190	4,500	125,000	4%	NIL		20.16
EPIKOTE 828 + VERSAMID 125	2,180	6,900	112,000	7.7%	1.6%	6.25%	47
EC2216	1,920	1,600	3,100	50%	NIL	5.7%	71.5
EC3559	5,450*	3,600	94,000	3.8%	NIL	5.7%	10.64

TABLE 5
COMPARISON OF ADHESIVES BY VARIOUS MECHANICAL & PHYSICAL PROPERTIES - ROOM TEMPERATURE DATA

SHEET 2 of 2

ADHESIVE	LAP SHEAR STRENGTH ON ANODISED AL.ALLOY	TENSILE STRENGTH OF RESIN	TENSILE MODULUS OF RESIN	STRAIN TO FAILURE TOTAL	PLASTIC STRAIN AT FAILURE	WATER SOLUBILITY COEFFICIENT	STRAIN ENERGY AT FAILURE N.mm
AF 163	5,000	6,341	95,650	8.6%	1.8%		45.5
EC 3524	1,350	1,300	33,000	4%	.3%	35%	4.465
EPIKOTE 815 + RTU	783	10,000	178,000	6%	NIL	7.35%	42
EC3568	3,150	2,391	98,550	2.5%	NIL		4.125
EC3578	2,872	2,536	87,000	2.5%	NIL		4.55
PERMABOND E34	2,174	3,913	182,000	2%	NIL		5.4

* F.P.L. etch   (2) Test Specimen Number   (3) Test Specimen Number

Generally, cold-sets can achieve adequate strength but not stiffness. Additional layers of fabric are likely to be necessary to restore stiffness. S.M.Lee (Refs 17-20) equates the requirements for adhesive properties for a lap joint with those for a strong, tough composite. Maybe existing composite resins are too stiff. See Refs 21, 22 and 23.Ref.21 shows that above a certain yield stress of the matrix resin the compression strength of the composite is virtually constant.

Diffusion and solubility coefficients obtained so far are given in Table 4A. Tables 4B to 4F give some data from the literature for comparison. It can be seen that water diffusion rates and equilibrium values at saturation vary quite considerably for both cold-sets and hot-sets.

Resin hardness, dry and wet after various cure temperatures is given in Table 6 (Courtesy of RAE). Unfortunately the hardness range fell between two scales for which test equipment was available. The work will be repeated when equipment for a more suitable scale is obtained. Table 1 gives T_g dry after various cure temperatures. Wet T_g data will have to be reported at a later date.

DISCUSSION

The initial results have shown, Table 4A, that water uptake varies considerably from one adhesive to another and in the case of EC2216 that a high initial rate of absorption (high diffusion coefficient) does not necessarily mean a high total water uptake. It follows that it is necessary to obtain Diffusion and Solubility Coefficients for each adhesive system under consideration for repair work and that the solubility coefficient may be the more important factor for long service lives, especially where metal joints are concerned.

Total moisture uptake in the service environment is important because of its effect on T_g and modulus for composites and because, above a critical value, it governs the rate of debonding of steel/epoxy joints, Ref.2., (Kinloch) and by implication that of aluminium alloy and other metal joints. Service experience with aluminium alloys and adhesives having a high water uptake has confirmed this. Unfortunately a unique critical water uptake cannot be defined because it varies with the adhesive, adherend and surface preparation combination selected.

TABLE 6

| ADHESIVE | HARDNESS DRY & WET AFTER VARIOUS CURE TEMPERATURES | | | | | |
	RT Cure Dry	RT Cure Wet	50°C Cure Dry	50°C Cure Wet	80°C Cure Dry	80°C Cure Wet
EA 9330	NR	90 SH	NR		NR	
EPIKOTE 815 + RTU	16 B	NR	12 B		22 B	
EPIKOTE 828 + VERSAMID 125	NR	NR	NR		NR	
EC 2216	NR	95 SH	NR		NR	
EC 3524	NR	90 SH	NR		NR	
EA 9321	3 B		21 B		21 B	
AF 163 *	---	----	---	---	NR*	
EC 3559	NR		NR		NR	
EA 9330 + 20% MICROBALLOONS						
EC 3568	17 B		---	---	---	---
EC 3578	NR		16 B		NR	
PERMABOND E34	22 B		13 B		27 B	

* AF 163-2M film adhesive cured at 120°C for 1 hour

NR (No Reading) means too hard to register on Shore "A" scale and too soft to give a reading on a Barcol Hardness Meter.

SH = SHORE
B   = BARCOL

The choice of base resin and curing agent has a marked effect on moisture absorption. Ref.3 (J.L.Tegg) states "For the sorption of water by polar polymers, polar groups such as -OH, -COOH, -NH$_2$, may act as specific sorption sites, and equilibrium sorption may depend not only on the quantity and nature of the polar groups, but also their positions on the polymer chain". Danieley, Ref.4, states that water uptake increases with degree of cure because of the increase in the number of hydroxyl sites. Table 4A shows that, in most cases, the water uptake is reduced by increasing the temperature of cure. From Tables 4B to 4F and the literature it would seem that chemistry can be adjusted, to some extent, to reduce water absorption. However, if polar groups are helpful to adhesion then the "best" adhesives may be those that take up the most water. Refs.5, 6. Where long term adhesion is required it may well be better to start with less adhesion and retain it, than to start with more adhesion and lose it. For adhesion to carbon-fibre, where water does not displace the adhesive at the interface, the "better" adhesive that absorbs more water may not be too much a of a problem. Fig.12 suggests that this is so. In this figure carbon-fibre lap shear test pieces, bonded with two different cold-setting epoxies are compared for strength after water immersion at room temperature. For aluminium alloy joints, experience with FM 1000, a nylon/epoxy with a high water uptake, suggests that a resin with a low total water uptake is the better choice. More recent experience with Hysol EA9330 confirms this.

The only time that airline aircraft are likely to be at temperatures above T$_g$ is on the ground in hot climates. At present the parts made from composites are under virtually zero load under these conditions.

However, when wings, tailplanes, fins or fuselages are made from composites the repair of upper surfaces with cold-sets could be a problem. The questions are how far T$_g$ is reduced by water uptake and how significant is this? Strength v temperature curves for various adhesives indicate that performance is acceptable for a few degrees above T$_g$ at least for short periods. Data giving compression properties is particularly necessary for upper wing surfaces. Compression data can be very different from tension data. See Figs.11B and 11C. Most large aircraft flight time is spent at -20°C or below.

Some of the cold-setting systems tested so far showed very little plastic strain at failure. Epikote 828 + Versamid 125 failed at 1.6% plastic strain, EA 9330 at 1% but Epikote 815 + RTU, Permabond E34, EC3524, EC3559, EC3568 and EC3578 were completely brittle and showed virtually zero plastic strain at failure. 3M-AF163-2M, (a 120°C cure film adhesive) failed at 1.8% plastic strain. However, study of the data obtained in the test programme suggests that total strain at failure correlates better with lap shear strength of aluminium alloy joints than plastic strain. Total strain is the sum of the elastic and plastic strains.

The aim of this work was extended when it seemed likely to be interesting to plot the lap shear strength on anodised or F.P.L etched aluminium alloy against each of the material properties obtained in the test programme. Consequently, lap shear strength to ASTM-D-1002 was plotted against tensile strength of the resin Figs.3A and 3B, tensile modulus Fig.4, shear modulus Fig.5 (calculated from tensile modulus), elongation at failure Figs.6A and 6B. Tensile strength of resin ÷ √Shear modulus Fig.7, "T" peel strength Fig.8, and Strain Energy at failure in Fig.9. "T" peel strength is plotted against elongation at failure of the resin in Fig.10. Fig.11 shows tensile strength of resins v tensile modulus. Ref.7 (Adams & Wake) confirmed that the results were largely as expected although they have not been reported in this form before.

The linear relationship to adhesive tensile strength (except for Epikote 828/Versamid 125 and Epikote 815 + RTU) was remarkable, Fig.3A. Similar results were obtained with data for "Cybond" resins published by American Cyanamid, Fig.3B.

The relationship to tensile modulus was very interesting, Fig.4, and Fig.5 showed modern adhesives as high modulus but tough and the older brittle adhesives as having a greatly reduced strength in lap joints in spite of a higher modulus, which no doubt raises the stress concentration factor at the overlap ends. Fig.7 from Table 3 plots the lap shear strength against adhesive tensile strength ÷ √Shear modulus (Volkersen factor). This graph readily separates adhesives with a low strain at failure from those with a higher value. 3M-AF163-2M, a hot-setting film adhesive is obviously on a different curve from the cold-sets. The relationships to total strain at failure in Fig.6A, and strain energy at failure Fig.9 were also very interesting. An optimum strain to failure was indicated in both cases. Fig.6B, also from "Cybond" data sheets from American Cyanamid showed a different shape of curve for toughened adhesives.

Fig.11 shows a relation between tensile strength and tensile modulus for the resins tested. The effect of elongation at failure or toughness is again demonstrated. Fig.11A using data from Ref.15 shows that composite matrix resins generally have a much higher modulus than resins used for metal bonding. They also have much lower values of elongation to failure than those found necessary for aluminium alloy joints.

Figs.3A and 3B showed the effect of toughness and suggested that a fracture mechanics approach might be helpful. Data from Refs.8 & 9, giving Boeing wedge test results for some acrylic adhesives was plotted as lap shear strength on aluminium alloy joints versus fracture energy values from the Boeing wedge tests. These graphs Figs.13 and 14 showed that even with toughened acrylics there was an increase in lap shear strength with increasing fracture energy up to a plateau in much the same way as elongation in Fig.6B. Refs.8 and 9 were originally produced to compare the effect of surface preparation on fracture energy after one hour in water.

145

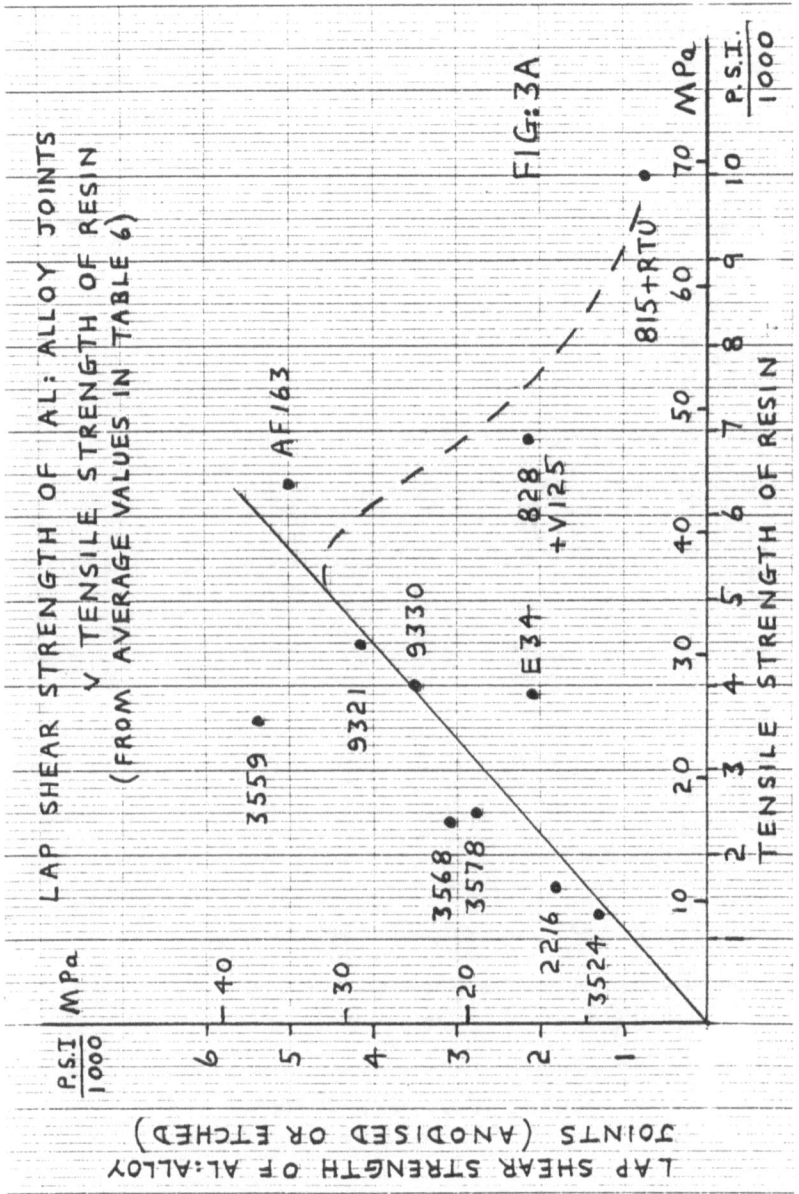

FIG: 3A

146

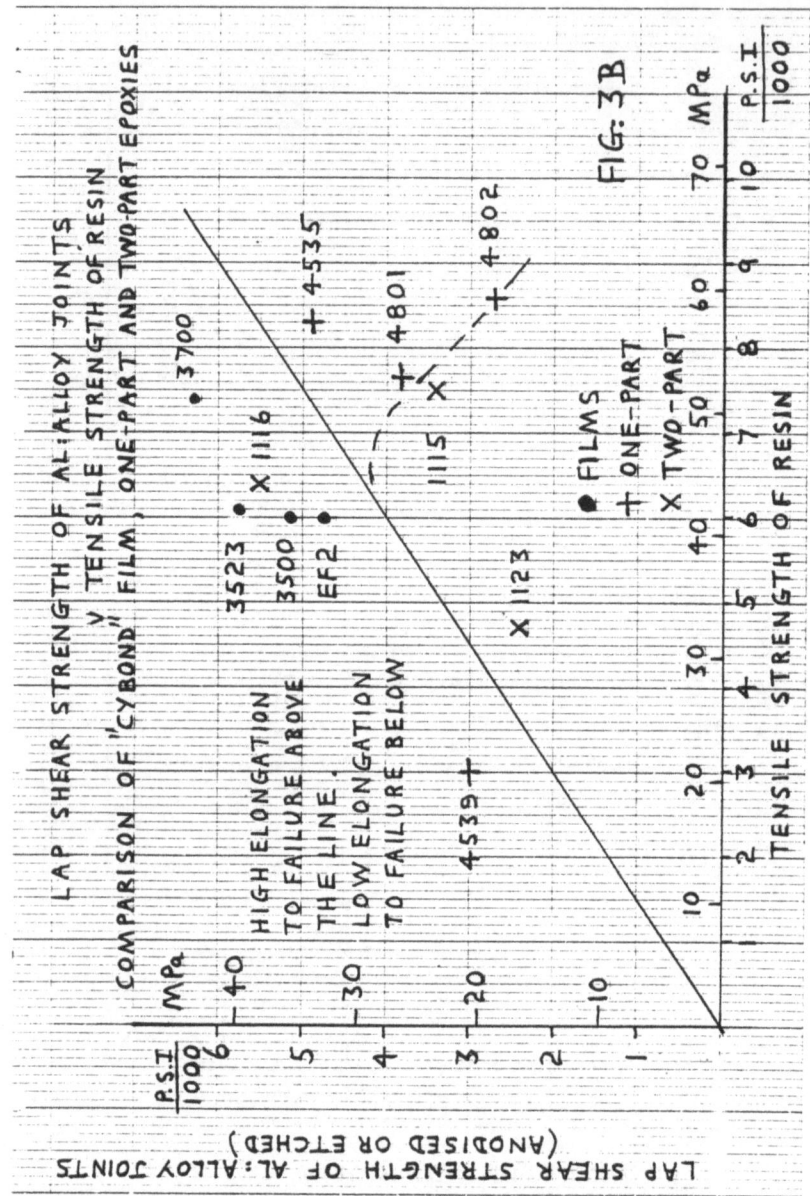

FIG. 3B

147

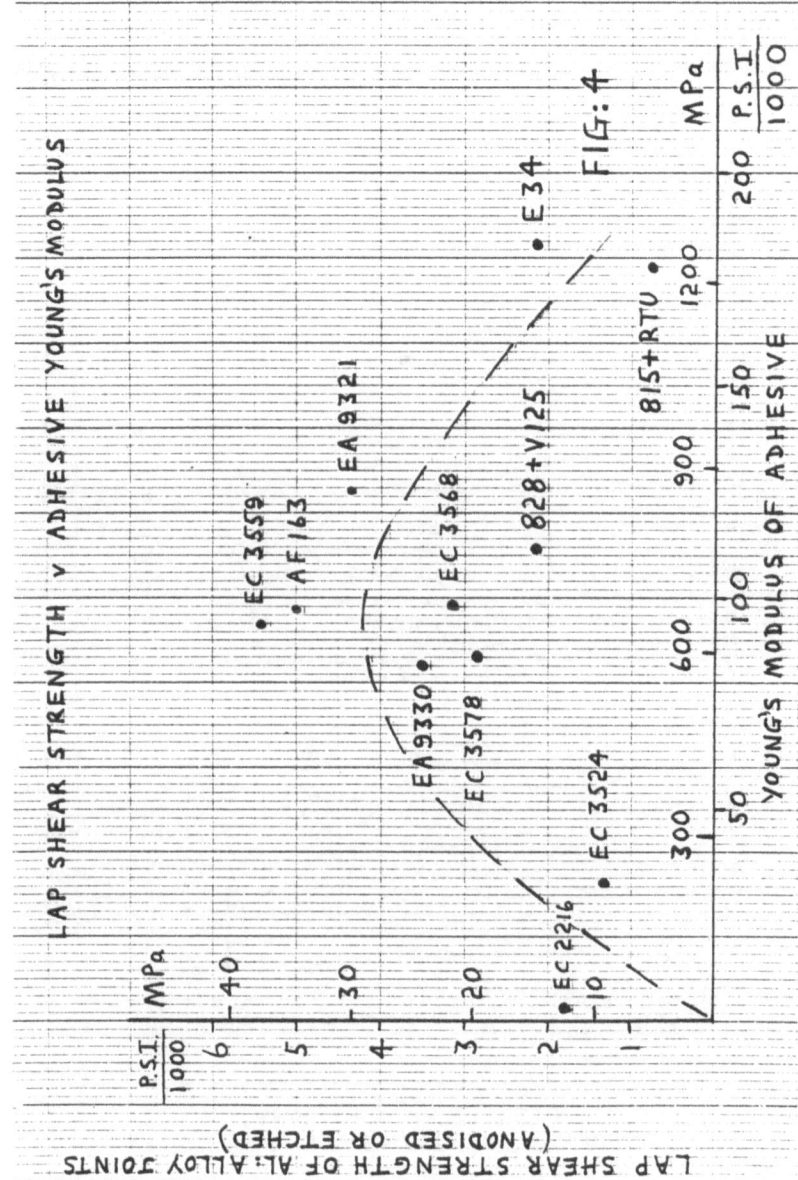

LAP SHEAR STRENGTH v ADHESIVE YOUNG'S MODULUS

FIG: 4

LAP SHEAR STRENGTH OF Al. ALLOY JOINTS
(ANODISED OR ETCHED)

YOUNG'S MODULUS OF ADHESIVE

148

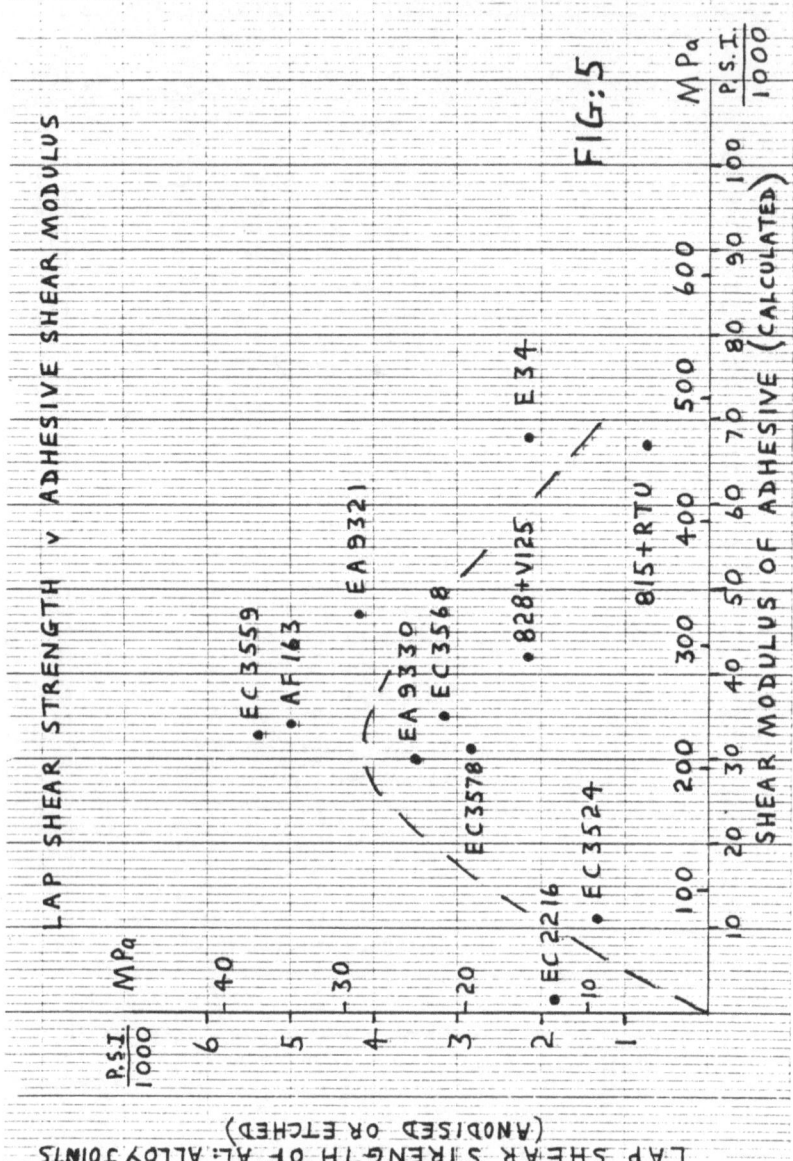

FIG. 5

LAP SHEAR STRENGTH v ADHESIVE SHEAR MODULUS

SHEAR MODULUS OF ADHESIVE (CALCULATED)

LAP SHEAR STRENGTH OF AL. ALLOY JOINTS
(ANODISED OR ETCHED)

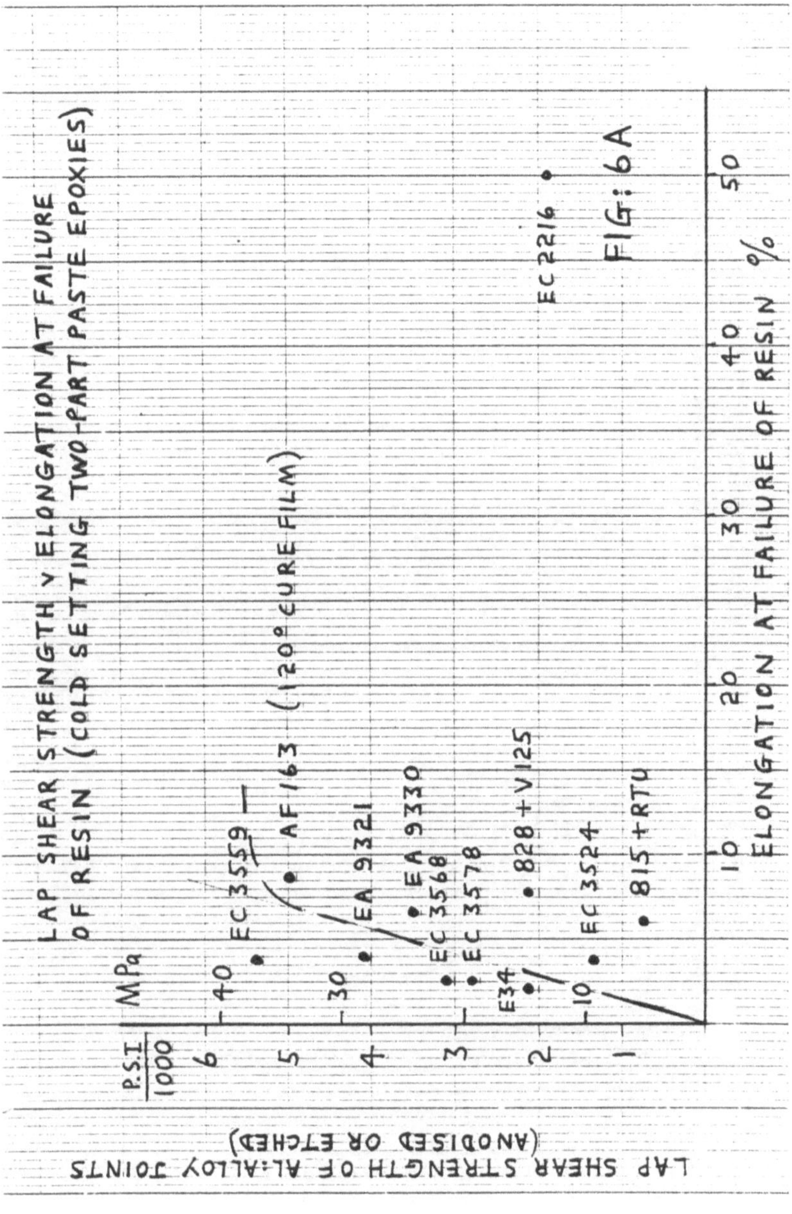

149

LAP SHEAR STRENGTH v ELONGATION AT FAILURE OF RESIN ((COLD SETTING TWO-PART PASTE EPOXIES)

AF 163 (120° CURE FILM)

EC 3559
EA 9321
EA 9330
EC 3568
EC 3578
E34
EC 3524
828 + V125
815 + RTV
EC 2216

FIG: 6A

MPa
40
30
10

P.S.I / 1000
6
5
4
3
2
1

ELONGATION AT FAILURE OF RESIN %
10
20
30
40
50

LAP SHEAR STRENGTH OF Al-ALLOY JOINTS (ANODISED OR ETCHED)

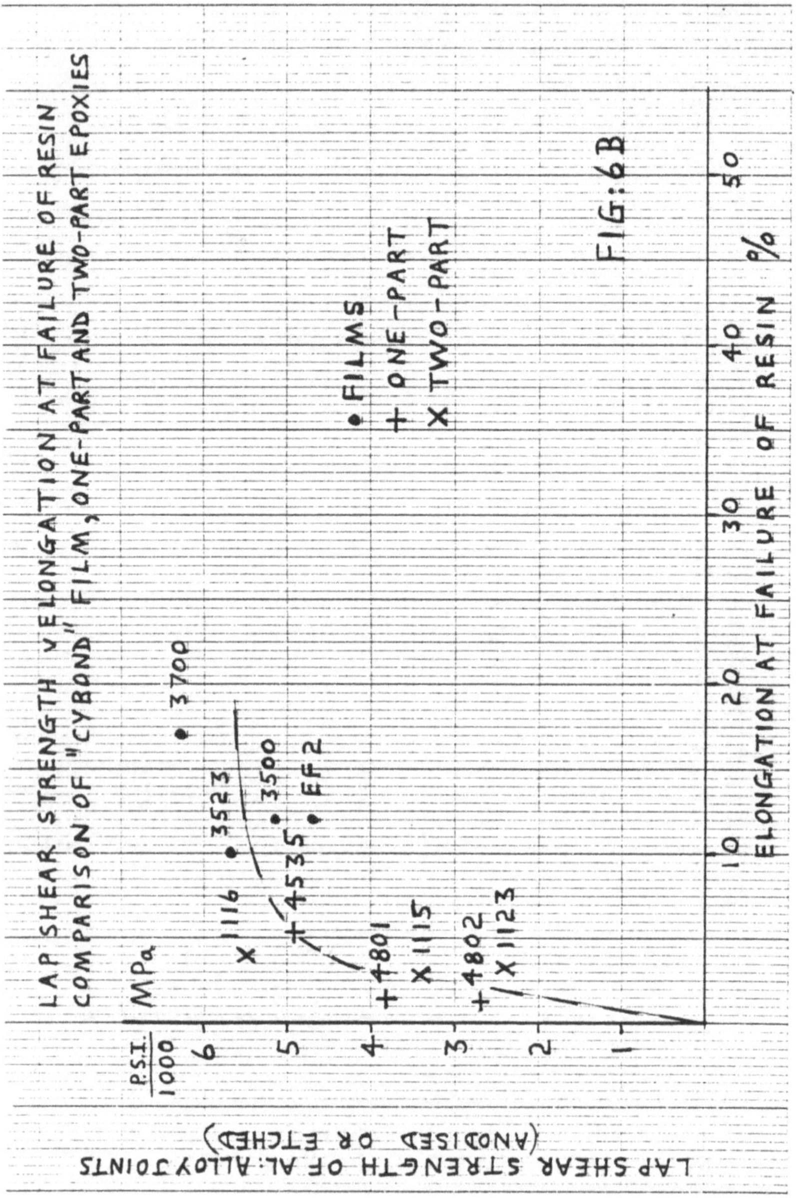
LAP SHEAR STRENGTH V ELONGATION AT FAILURE OF RESIN
COMPARISON OF "CYBOND" FILM, ONE-PART AND TWO-PART EPOXIES
MPa
P.S.I.
1000
6
5
4
3
2
1
3700
3523
3500
EF2
4535
116
4801
1115
4802
1123
• FILMS
+ ONE-PART
X TWO-PART
FIG: 6B
10
20
30
40
50
ELONGATION AT FAILURE OF RESIN %
LAP SHEAR STRENGTH OF AL: ALLOY JOINTS
(ANODISED OR ETCHED)

151

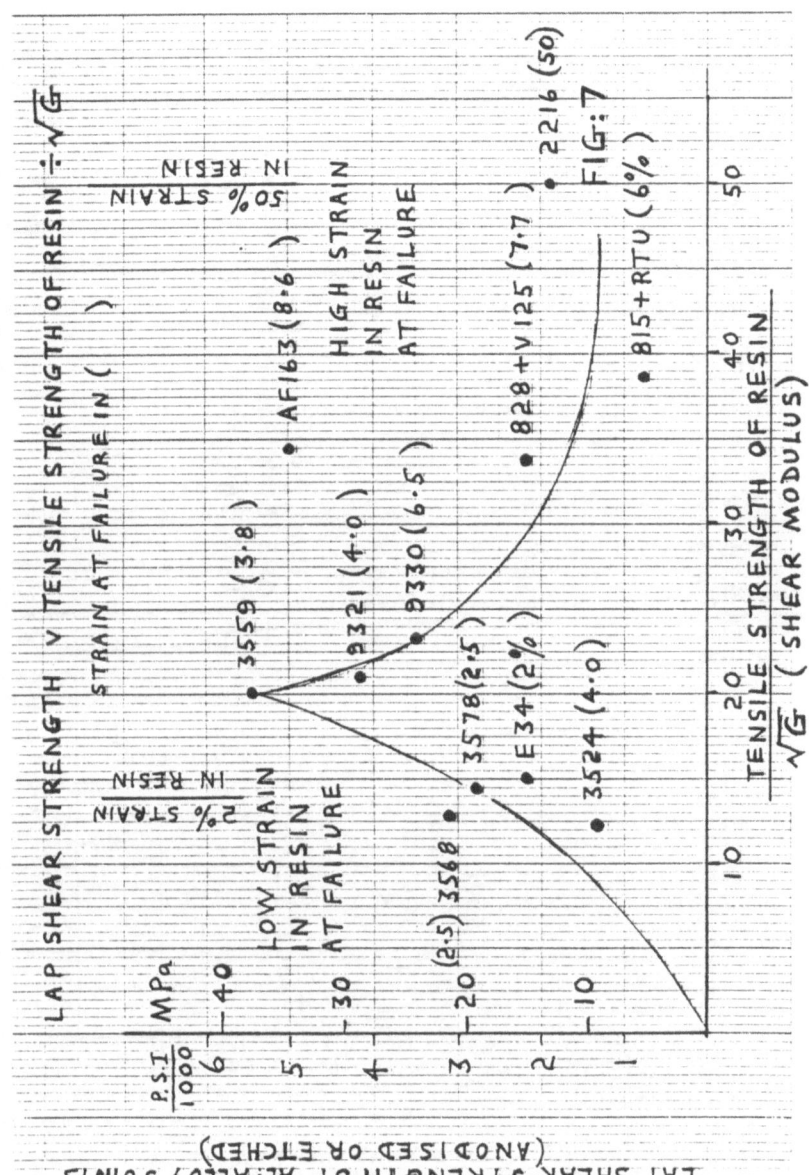

FIG:7

LAP SHEAR STRENGTH OF AL. ALLOY JOINTS
(ANODISED OR ETCHED)

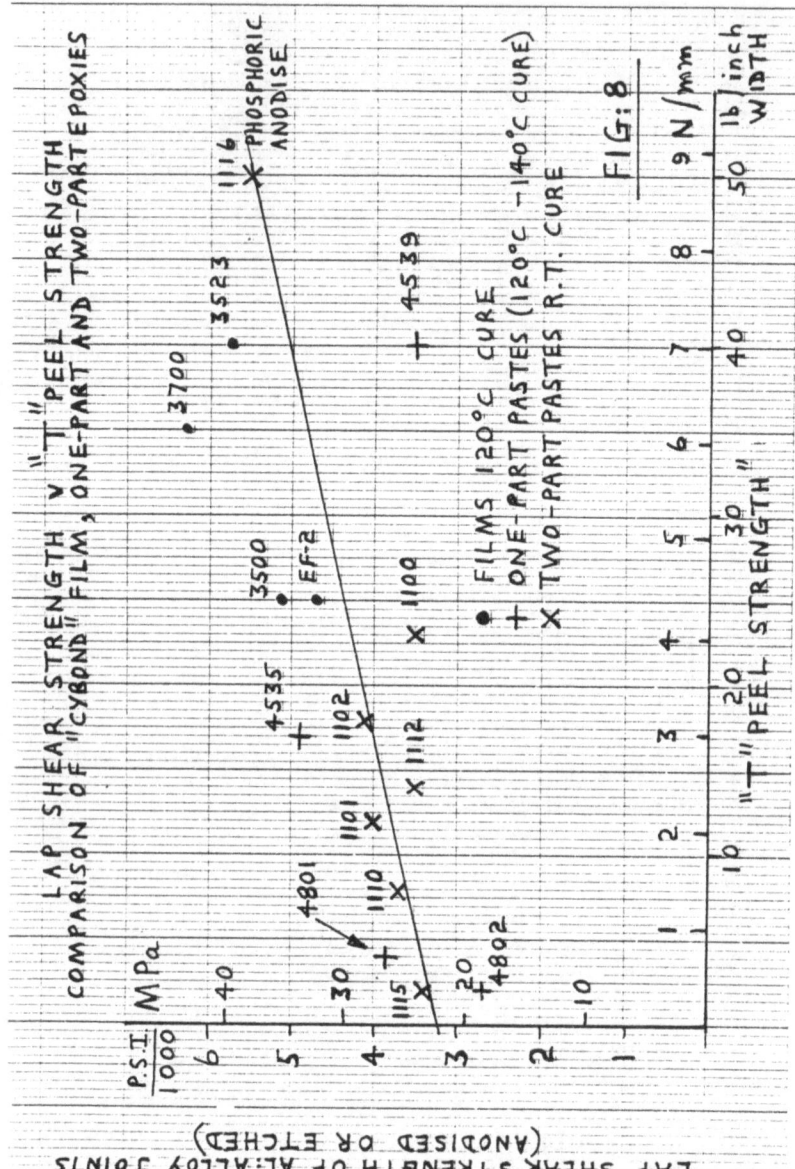

COMPARISON OF "CYBOND" FILM, ONE-PART AND TWO-PART EPOXIES

LAP SHEAR STRENGTH V "T" PEEL STRENGTH

FIG:8

"T" PEEL STRENGTH

LAP SHEAR STRENGTH OF AL-ALLOY JOINTS
(ANODISED OR ETCHED)

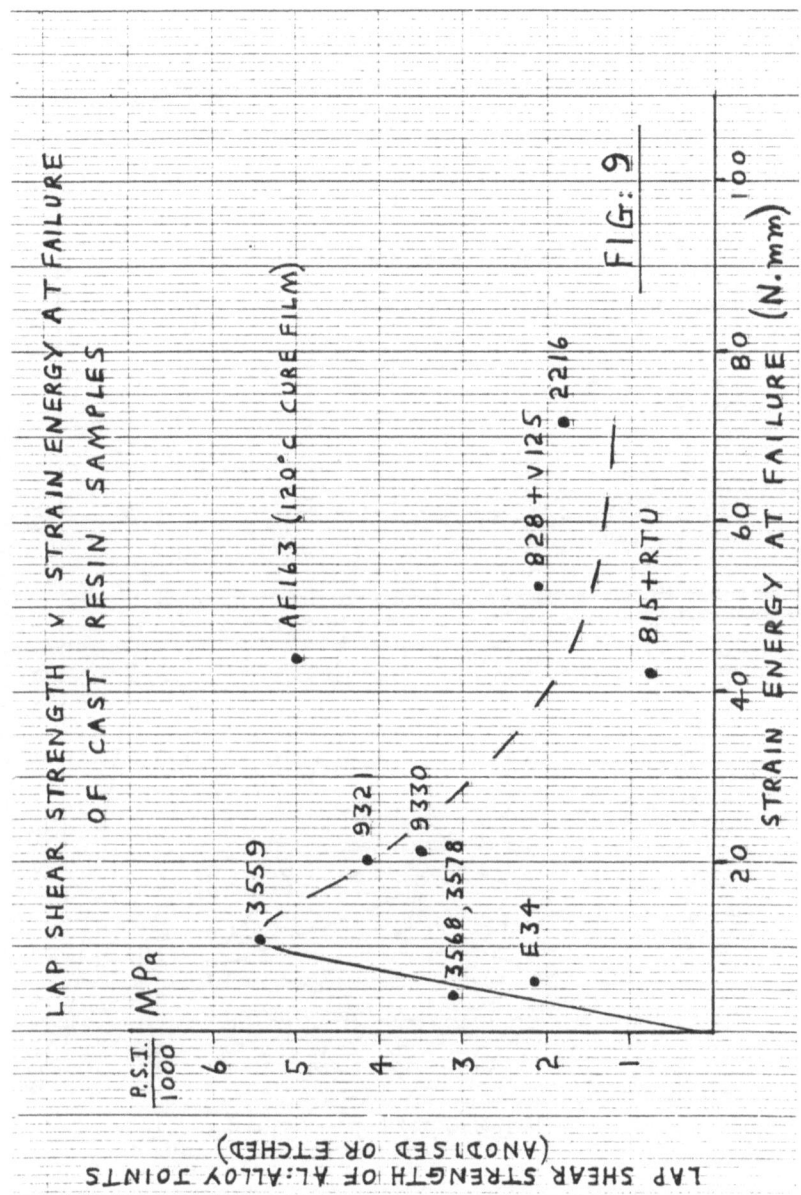

FIG: 9

154

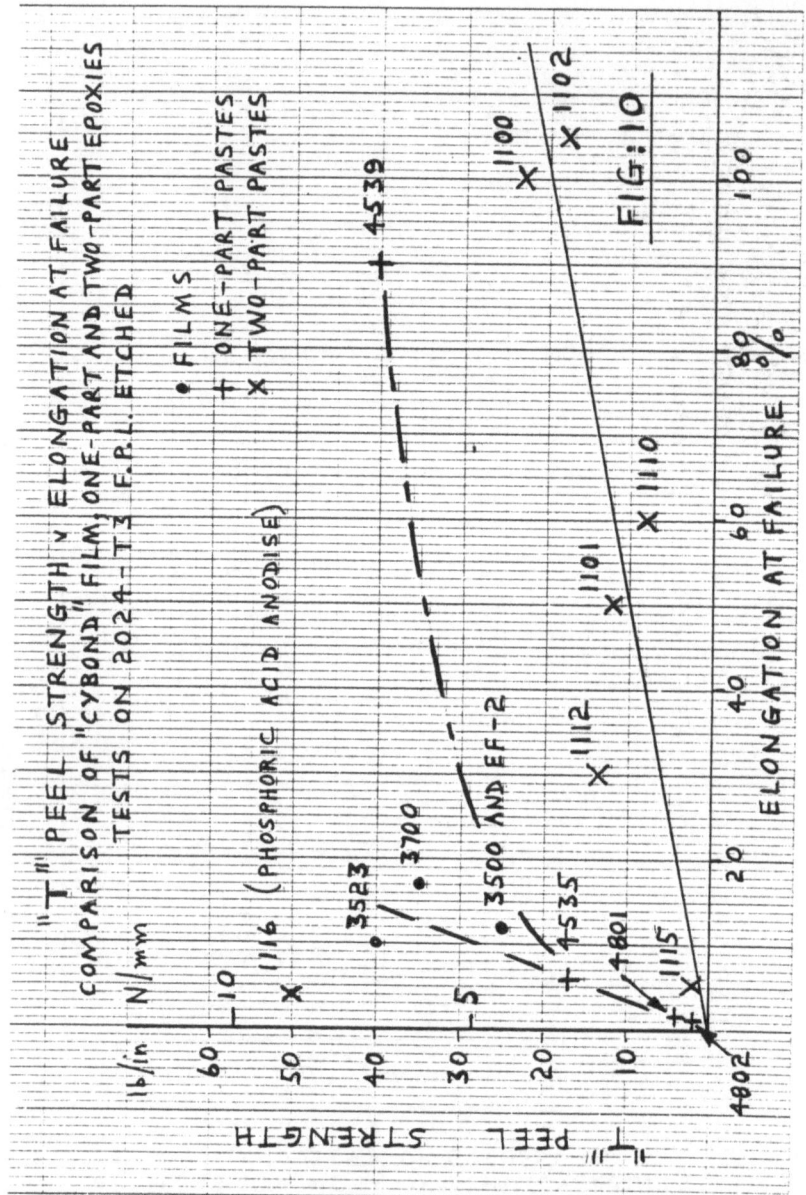

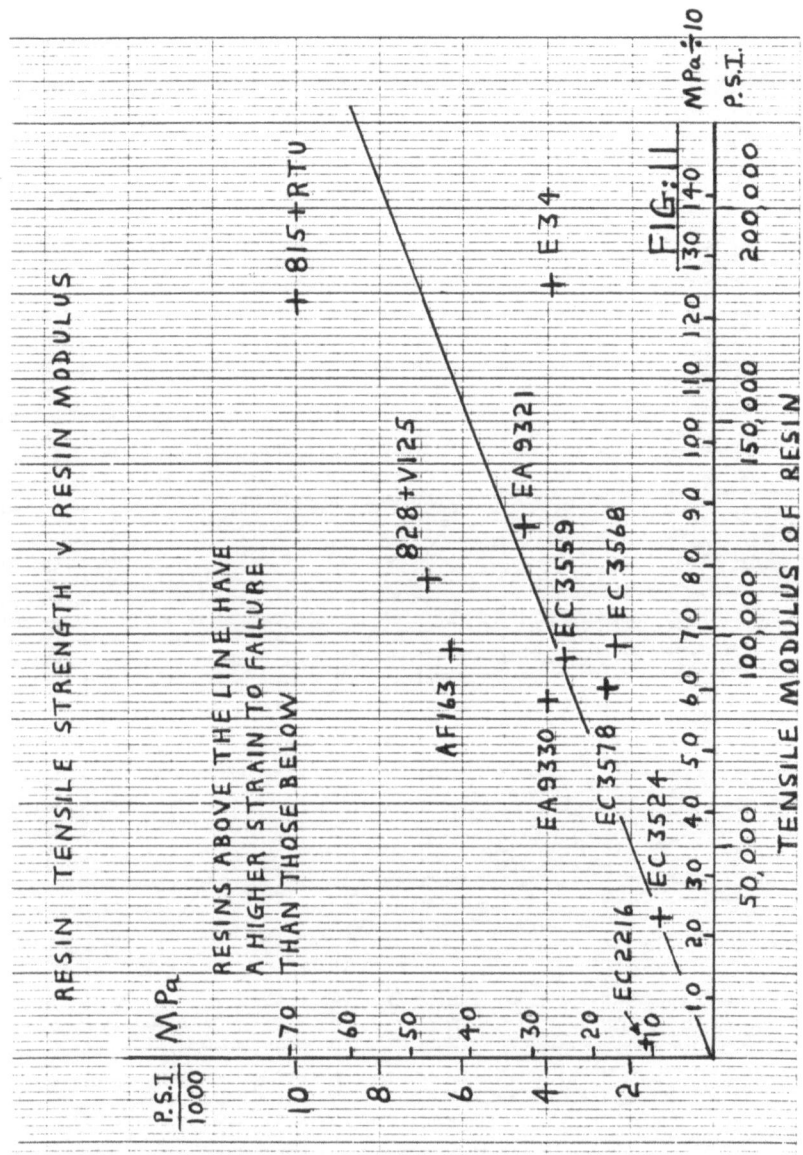

RESIN TENSILE STRENGTH v RESIN MODULUS

RESINS ABOVE THE LINE HAVE A HIGHER STRAIN TO FAILURE THAN THOSE BELOW

FIG. 11

TENSILE MODULUS OF RESIN

TENSILE STRENGTH OF RESIN

156

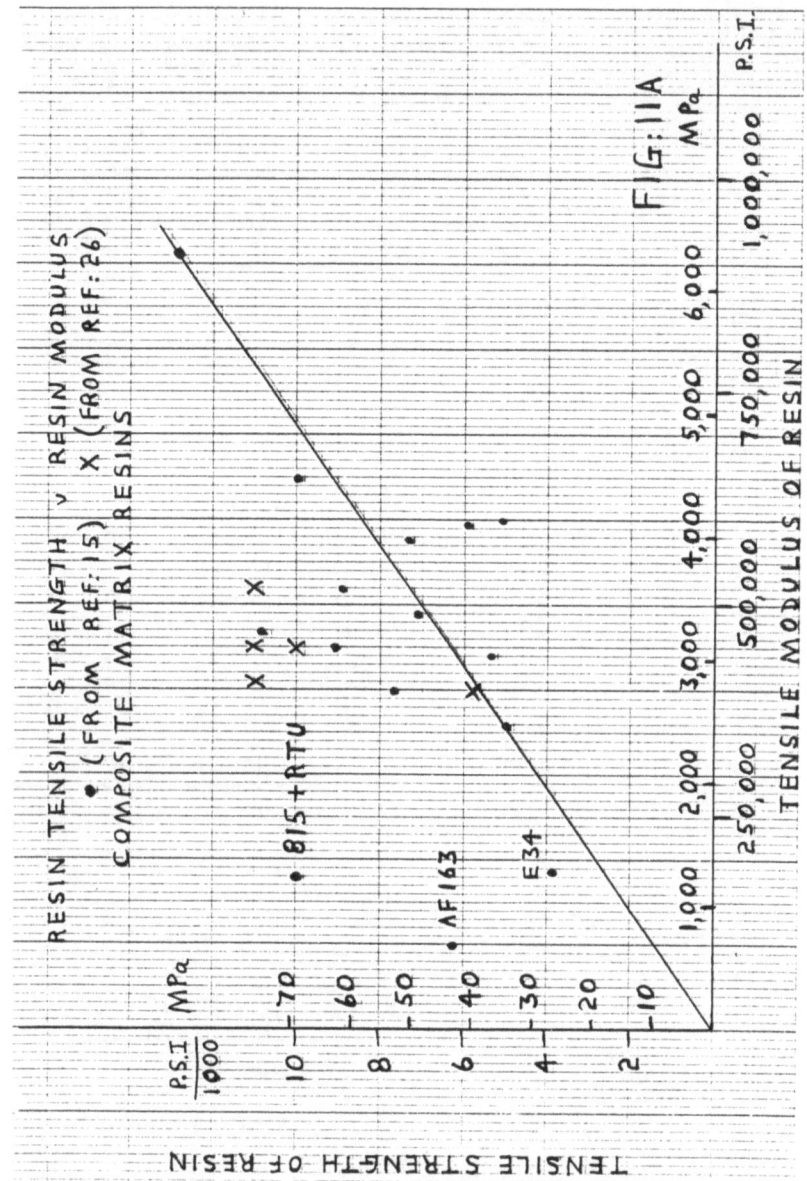

FIG:11A

157

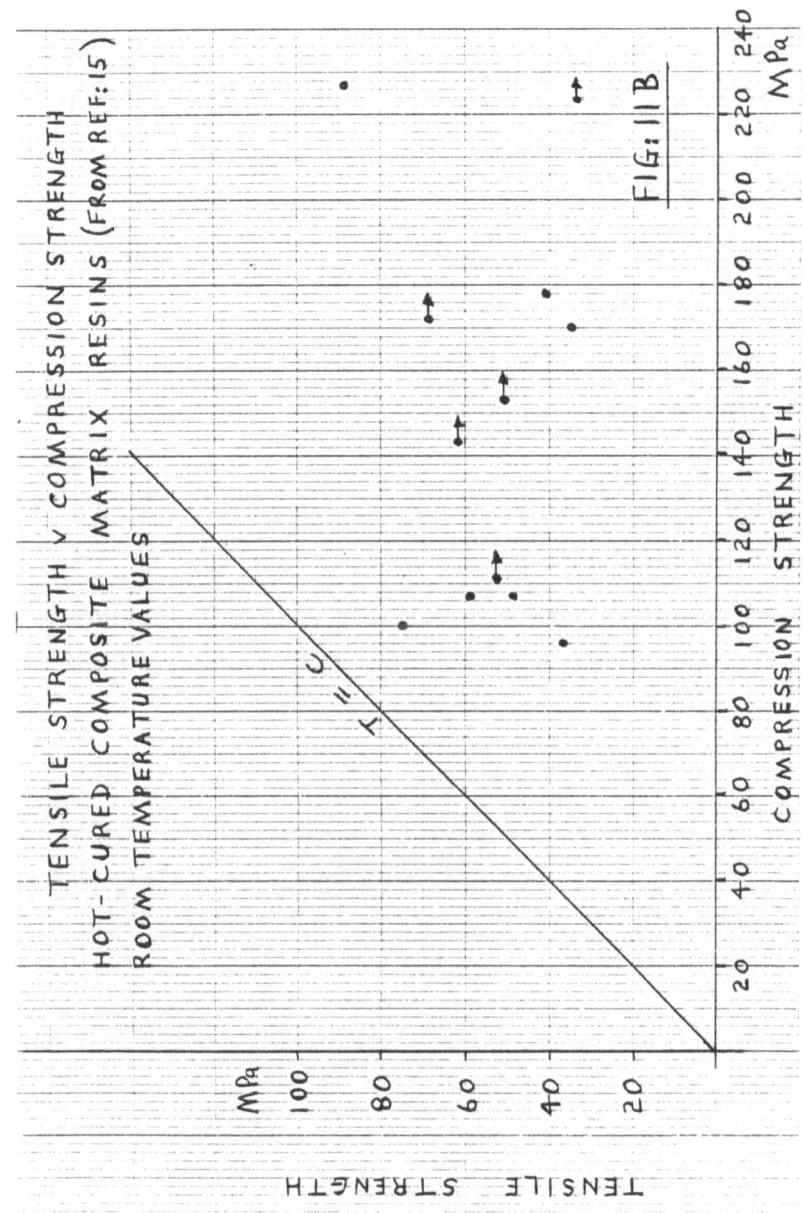

FIG: 11 B

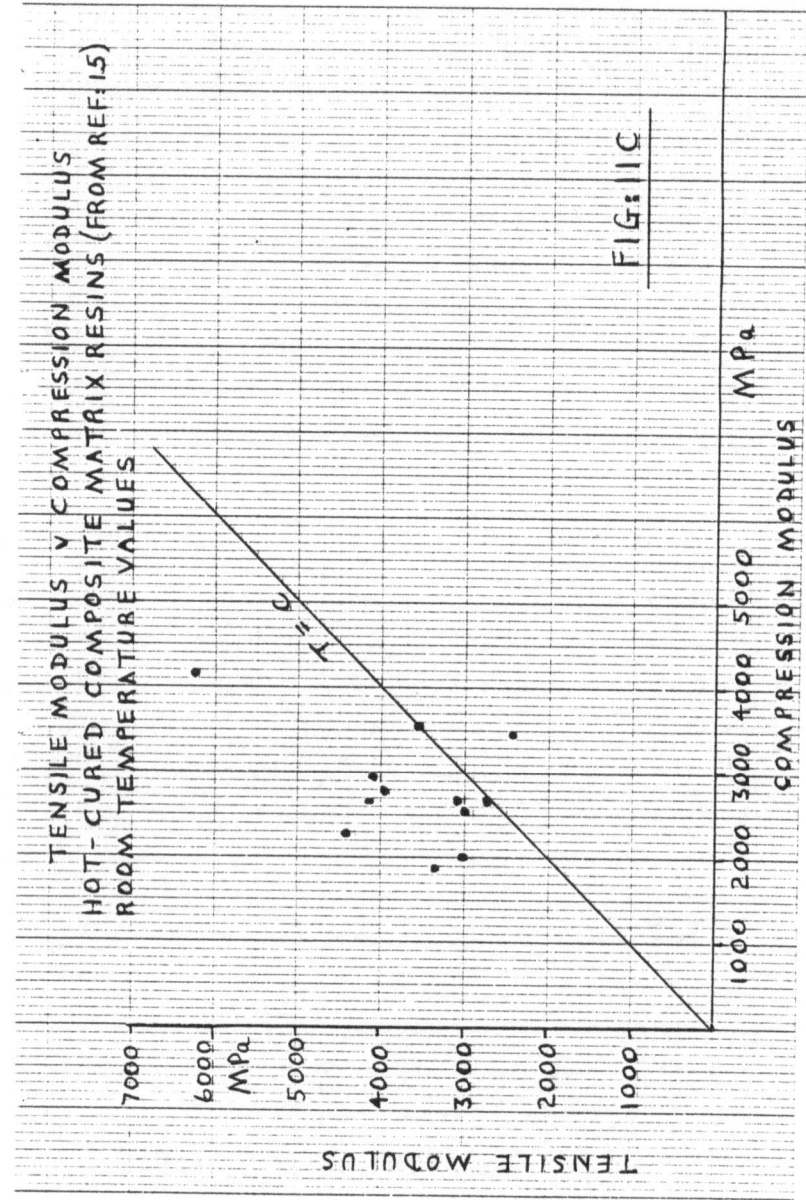

TENSILE MODULUS V COMPRESSION MODULUS
HOT-CURED COMPOSITE MATRIX RESINS (FROM REF.15)
ROOM TEMPERATURE VALUES

FIG: 11C

Unless the failure is fully cohesive the fracture energy obtained must be some function of the surface preparation. Fig.13 shows that the fracture behaviour of the resin interacts with the surface preparation and the best results can only be expected when failures are partly or wholly cohesive. Fig.10 shows Adhesive 1116 giving an excellent result with a phosphoric anodise surface preparation. Surface preparation seems even more important in peel and cleavage modes than in shear.

However, a fracture energy measurement obtained from a sample of the adhesive alone would be expected to correlate with fracture energies obtained in a wedge test. Further work on these lines could be useful.

The value of the lap shear v modulus curves, Figs.4 and 5 is that they clearly indicate that above a certain modulus joint strength falls again and that good adhesives can be made with fairly high modulus and therefore good creep resistance and an optimum in joint strength. However, if adhesives are given too high a modulus, by curing above the recommended temperature by the ommission of rubber toughening or by choice of curing agent, this is likely to give a reduced lap joint strength.

Ref.10 (Delmonte) suggests an optimum bond strength for composites.

Fig.11A suggests that if repair resins are used having a lower modulus than the composite matrix resin then some additional fabric layers may be necessary to restore the original stiffness. Ref.15 shows that Compression Modulus can differ widely from tension modulus and both sets of data are necessary. See Figs. 11B and 11C.

Fig.12, carbon fibre joints, shows that the joint strength for the stronger and stiffer Epikote 815 + RTU is lower than the joint strength for the weaker and lower modulus EA9330. It is worth noting that EA9330 is near the optimum position on Fig.4.

J.D.Minford in Ref.2, indicates that toughness can be achieved with very little loss of modulus.

Comparison of the Cybond range of adhesives using the American Cyanamid data gave similar results to the above work. Figs.3B, 6B, 8 and 10 illustrate these materials.

In the time available for the preparation of this paper it was not possible to make a detailed study of theoretical bases for the correlations suggested. Refs 7, 17-22, 27 and 28 cover some of the findings of other workers.

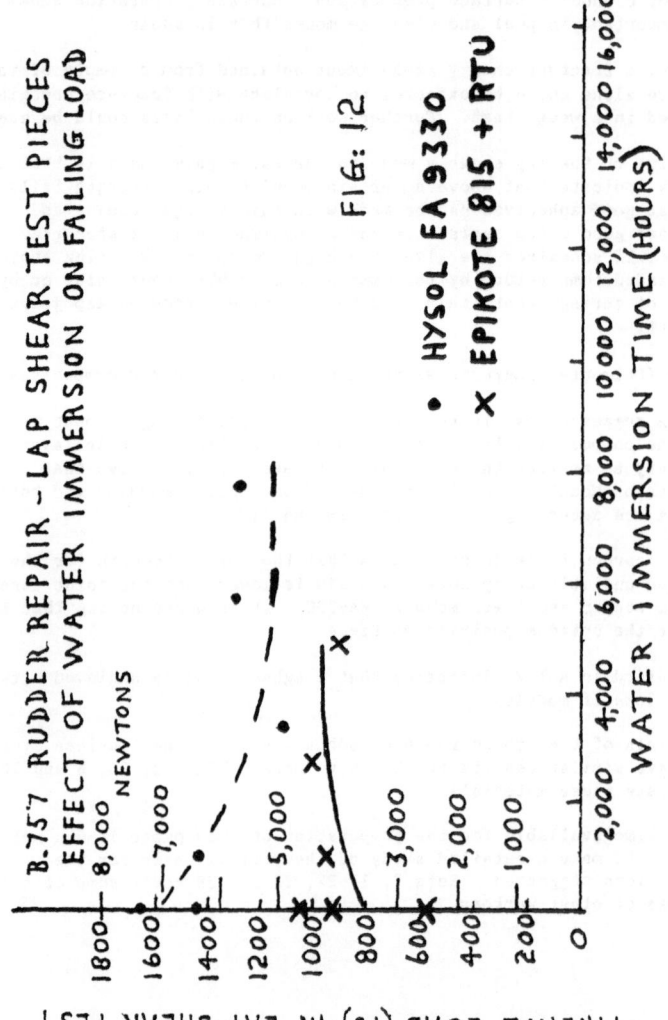

B.757 RUDDER REPAIR - LAP SHEAR TEST PIECES
EFFECT OF WATER IMMERSION ON FAILING LOAD

FIG: 12

• HYSOL EA 9330
✕ EPIKOTE 815 + RTU

161

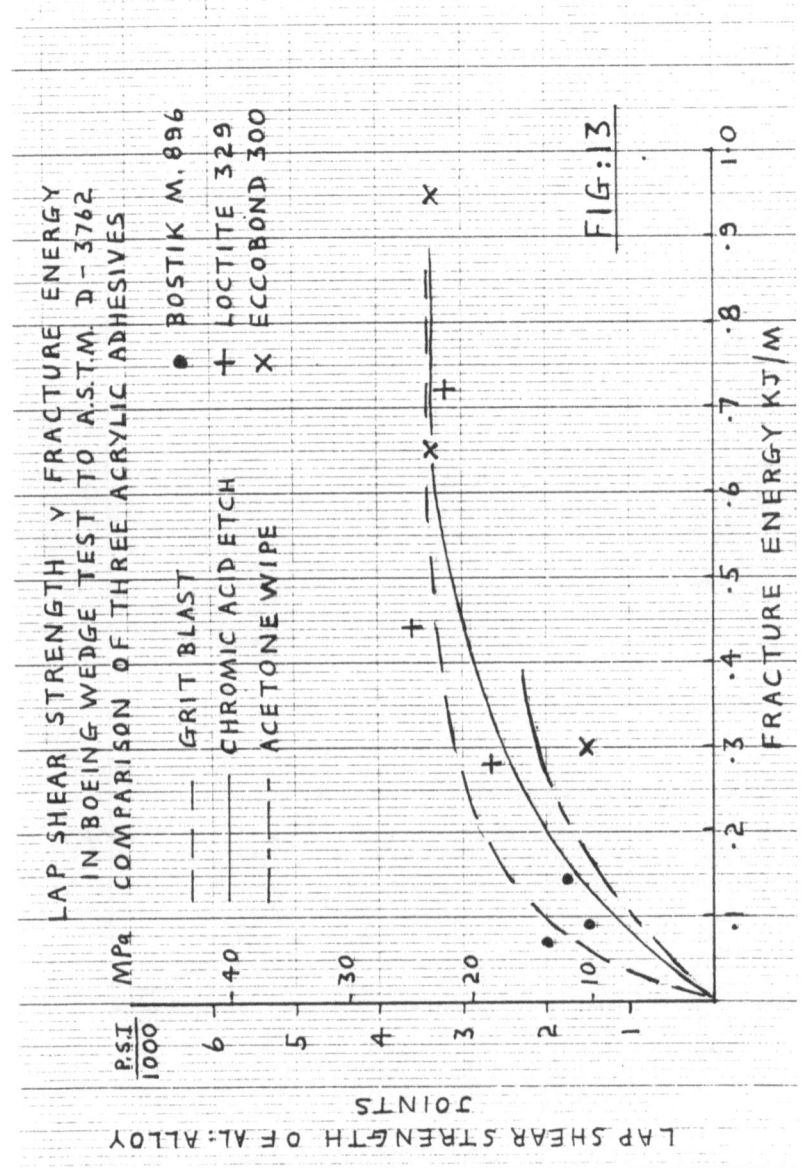

FIG:13

162

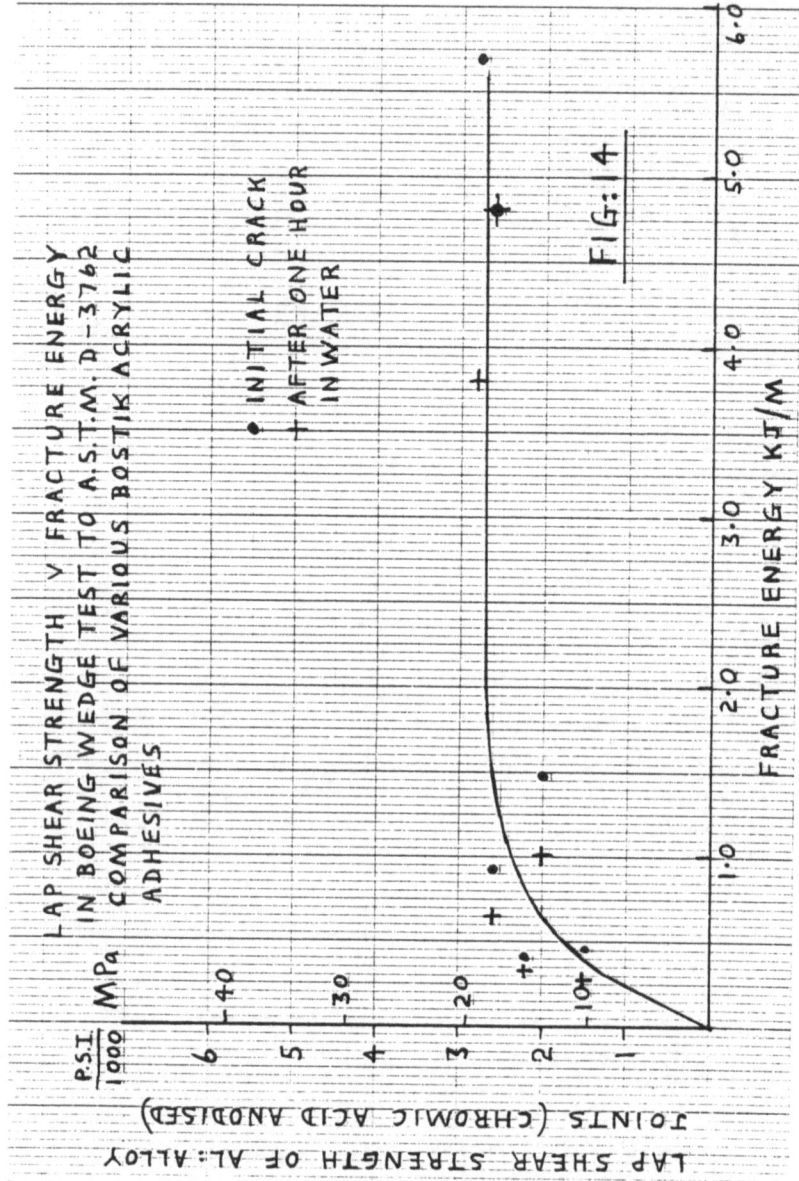

FIG. 14

DIFFUSION OF WATER INTO ADHESIVES AND JOINTS

Diffusion and Solubility Coefficients have been mentioned above. A
computer programme for the diffusion of water into metal lap joints was
adapted from the work of Anne Moloney et al, Ref.11, to operate on an IBM
PC. This enables the diffusion of water into a lap joint to be calculated
as a fraction of the total uptake for that particular resin.
Calculations were carried out for a standard 12.5 mm x 25 mm overlap test
piece to ASTM D-1002 to estimate time to saturation under total immersion
and also for a 38 mm x 250 mm joint to simulate a 1½" overlap commonly
used when bonding thin metal skins during aircraft repairs. Only in wet
areas of an aircraft would total immersion be realistic but Fig.15 shows
the time to various fractional water uptakes for different diffusion
coefficients at the centre of a 38 mm overlap joint and Fig.16 for an
12.5 mm overlap. Fig.17 shows the time for various levels of water
uptake at a section across the centre of a 12.5 mm joint for a diffusion

coefficient of 10^{-12} m^2. s^{-1}. Fig.18 shows similar information for
a 38 mm joint. When these figures are related to the Diffusion
Coefficients of commonly used adhesives it can be seen that in wet areas
it is possible to get a rather wet joint after only a few years.

For thin composites (2 mm or less) it must be assumed that equilibrium
uptake will occur in quite a short time, probably less than one year.
A figure of 1% is usually quoted for hot-cured pre-pregs although some
authorities are suggesting an increase to 1.3%. Ref.16. This reference
suggests ageing at 84% RH at room temperature to simulate the worst world
operating conditions. Ref.25 suggests that the proportion of the
equilibrium uptake at 100% RH or immersion,absorbed at other RH levels,is
very approximately pro-rata or slightly less than pro-rata.Therefore if
the equilibrium uptake by total immersion is known,a rough estimate of
the likely uptake at various humidity levels can be made. 1.3% as 80% of
equilibrium uptake, relates to an equilibrium uptake of the resin alone
of about 5%. Resins with lower values than this are available see Table
4.
Wright (Ref.13) suggests that Diffusion Coefficients for composites are
about one order of magnitude less than those for the neat resins.

There is a need to develop adhesives for composite and bonded metal
repairs having lower Solubility Coefficients, whilst retaining all the
desirable properties previously mentioned. Although Diffusion
Coefficients are important and low values would be helpful to slow down
the rate of moisture uptake, they would also reduce the speed of drying.
For thin composites, which will always eventually reach equilibrium, the
reduction of total uptake is probably much more important. This is also
likely to be true for bonded metal joints where corrosion or debonding
appears to take place when the water content exceeds some critical, but
not yet well defined, value. Moloney et al, Ref.11, states that the
strength of a joint has been correlated with its' fractional water
content.

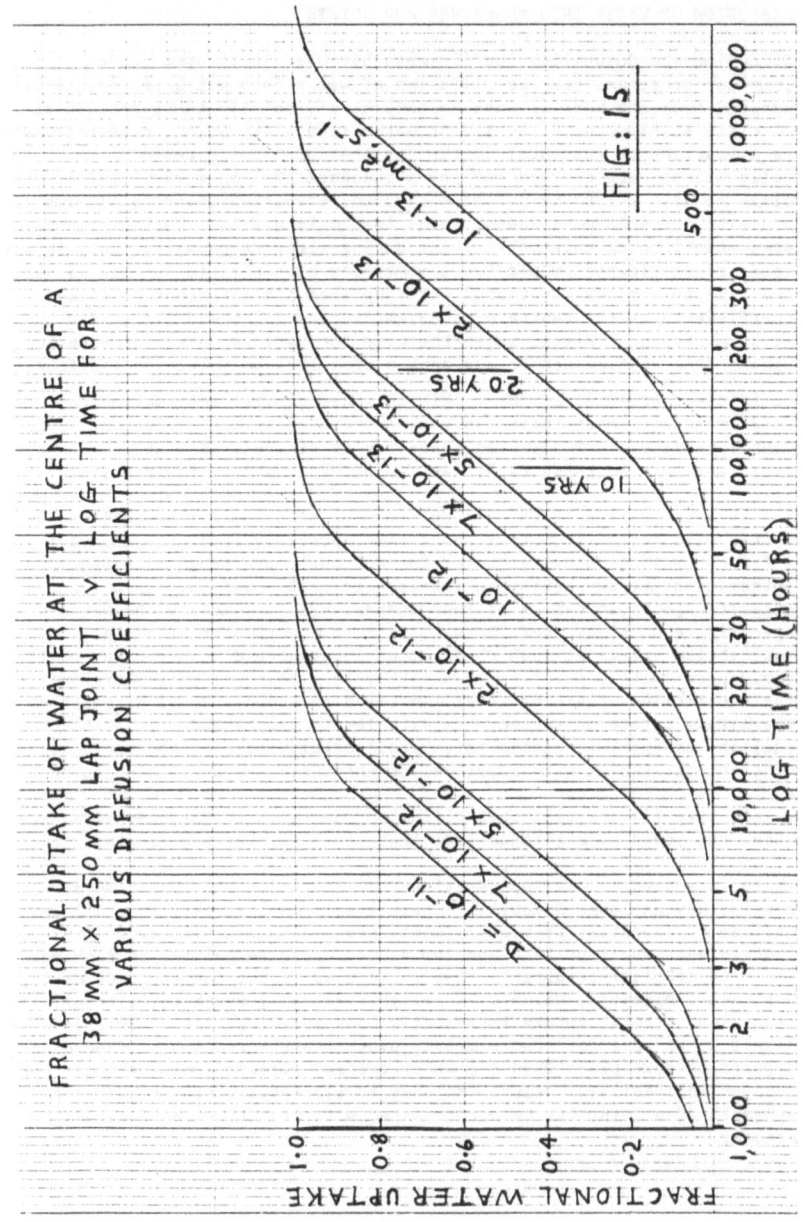

FRACTIONAL UPTAKE OF WATER AT THE CENTRE OF A
38 MM × 250 MM LAP JOINT v LOG TIME FOR
VARIOUS DIFFUSION COEFFICIENTS

FIG. 15

165

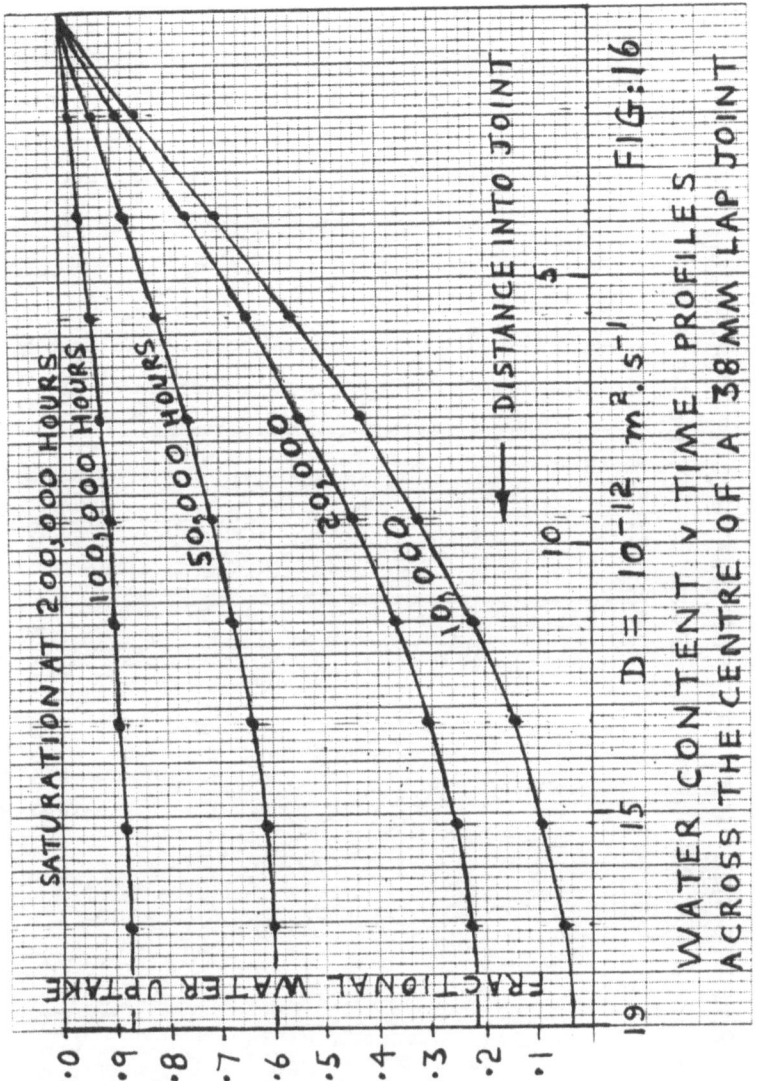

FIG:16

D = 10⁻¹² m².s⁻¹

WATER CONTENT v TIME PROFILES ACROSS THE CENTRE OF A 38 MM LAP JOINT

166

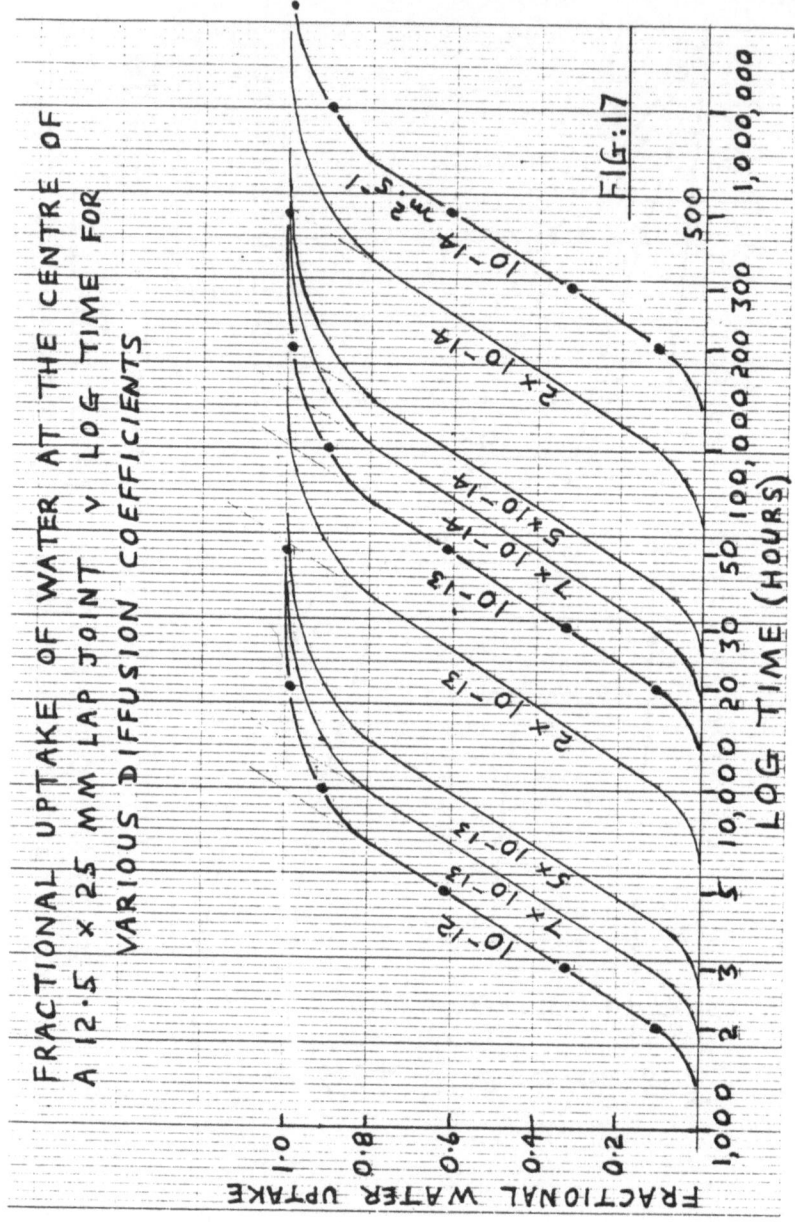

FRACTIONAL UPTAKE OF WATER AT THE CENTRE OF A 12.5 × 2.5 MM LAP JOINT v LOG TIME FOR VARIOUS DIFFUSION COEFFICIENTS

FIG:17

FRACTIONAL WATER UPTAKE

LOG TIME (HOURS)

167

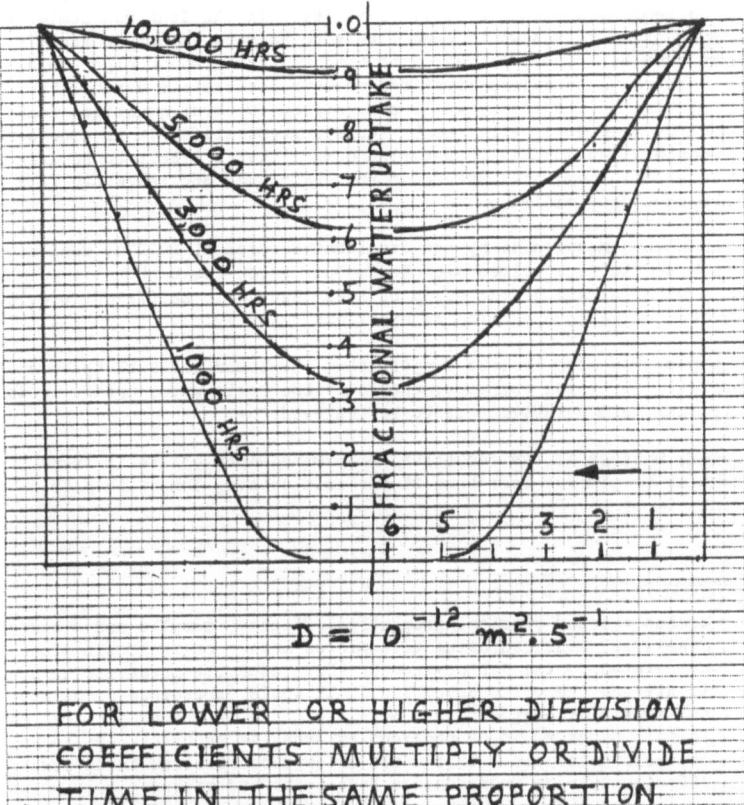

$$D = 10^{-12} \, m^2 . s^{-1}$$

FOR LOWER OR HIGHER DIFFUSION
COEFFICIENTS MULTIPLY OR DIVIDE
TIME IN THE SAME PROPORTION

FIG: 18
WATER CONTENT v TIME
PROFILES ACROSS CENTRE OF
12·5 MM LAP JOINT

Ref.13 states that the equilibrium moisture content depends more on
relative humidity than temperature and the diffusion coefficient depends
more on temperature than relative humidity. In the long run humidity in
the area of operation would seem likely to be more important than
temperature but tropical areas with high temperature and high humidity
have been shown to be the worst for moisture uptake and durability in
confirmation of the theory.While recognising, without question, that
durability is greatly affected by surface preparation, it is considered
that for a given surface preparation durability is likely to be affected
by the equilibrium water uptake of the resin used and therefore this
uptake should be minimised.

Fokker have demonstrated by the excellent service of EC2216, when used on
an anodised surface primed with Redux 101, a hot-cured phenolic primer,
that a water uptake of the resin of less than 5% is sufficient provided
that the surface preparation is excellent. Ref.29.

It is probably also true to suggest that the lower the quality of surface
preparation of metal surfaces, the lower the resin water uptake needs to
be if long term durability is to be achieved.

However, because of the reduction of T_g with water uptake it seems
essential to aim for less than 2% for cold-setting composite matrix
resins.

CONCLUSIONS

1.   Mechanical and physical properties of epoxy, acrylic and other resin
     adhesives, measured using test pieces made from the resin alone, can
     be related to the performance of those adhesive materials in bonded
     joints and composite assemblies. It would, therefore be helpful to
     the repair of composite and metallic parts if the properties listed
     in this paper could be provided by Adhesive Manufacturers to users
     of their materials. This would allow substitutions to be more
     effectively made when repairs have to be undertaken.

2.   Up to a limit, lap shear strength rises with adhesive tensile
     strength. Toughened adhesives are clearly better than others.

3.   There appear to be optimum tensile modulus and shear modulus values
     giving good joint strength and creep strength although toughening
     will affect the position of an optimum. The achievement of improved
     impact resistance for composites may mean using tougher and lower
     modulus resins.

4.   A good fracture energy in the Boeing wedge test is very desirable.

5. Figs.6A, 6B and 7 indicate that best joint strengths are obtained when the elongation to failure of the adhesive itself is about 5 to 10%. Joint strength falls quite sharply as the elongation to failure falls below 5%. Strength and modulus usually increase with higher post-cure temperatures but elongation to failure would be expected to fall.
This work could usefully be repeated at various post-cure temperatures and at the minimum and maximum temperatures expected in service. Ref.24. In general adhesives become softer at higher temperatures and harder and more brittle at lower temperatures. It could also usefully be repeated on different adherend materials as stiffer adhesives should be useable with stiffer adherend materials.

6. A good adhesive for repair purposes would seem to require the following properties when making bonded metal joints.

   i)   A tensile modulus of 90 - 100,000 psi when obtained using the specimen design and strain rate (.01) used for these tests at room temperature.

   ii) An elongation at failure of about 10%.

   iii) Adequate strength and toughness over the operating temperature range.

   iv) A T_g at equilibrium water uptake in the service environment suitable to the operating conditions. Say 60°C for subsonic aircraft.

   v)   A low Solubility Coefficient (less than 5%). Table 4A indicates that values below 2% are possible.

   vi) A low Diffusion Coeff. (less than $5 \times 10^{-13} m^2.s^{-1}$ at 20°C).

7. T_g can be increased by warm curing to a useful extent.

8. T_g can be related to resin hardness to some extent.

9. Cold-setting resins can be used for repairs to composites and bonded metal structures but careful selection is required in relation to the service temperature over which the part will be used. Ref.14.

   Figs.11 and 11A show that composite matrix resins generally have a tensile modulus two or three times greater than resins used for metal joints. Stiffer resins need to be used for composite repairs than for metal repairs. Compression data is necessary for the selection of repair adhesives for composites. For composite repairs additional fabric layers may be necessary to compensate for the reduced modulus of cold-set repair resins or interleaved film adhesives.

   Matrix dominated properties, e.g. flexural strength and stiffness, compression buckling resistance and impact resistance need to be related to fundamental matrix properties so that the effect of using a different matrix can be accurately assessed. Refs 17-24.

The value of T_g in the wet state and any corresponding reduction in strength may be the determining factor in the use of cold-setting adhesives (even after warm curing) for some repair applications. Aircraft manufacturers require a significant degree of strength retention at 80°C (180°F). Lower water uptake would give a smaller T_g reduction.

10. Strong but tough resins are recommended rather than hard and brittle resins. The repair of composites could possibly use a tougher resin for the first layer where a joint is being made and a harder and stiffer resin for the outer layers. A tough resin could also be used to bond pre-cured patches. Figs.6A and 6B indicate that the best cold-sets can approach the mechanical performance of hot-sets. However, if the total water uptake of a cold-set is greater than that of the hot-set it replaces, then its' service life will be shorter especially in a bonded metal joint and its' stiffness will be less in a composite.

   Of the cold-setting adhesives tested so far, EA 9321 would seem to be the best for bonded metal joints and Epikote 815 + RTU the best for composite repairs both mechanically and for moisture uptake.

11. Having drawn the above conclusions it must still be said that the selected adhesive must have good adhesion properties and good surface preparation has been shown to be even more important if cold-setting adhesives are used. It must also be resistant to jet fuel, hydraulic fluids, de-icing fluids, engine oils and any other chemicals found in the service environment.

   Practical factors such as viscosity, for ease of application, and pot life, which governs the size of job that can be done, are also very important. Low toxicity is also important for repair work, which may be carried out in a hurry when production line safety precautions are not available.

ACKNOWLEDGEMENTS

The assistance of Dr.W.W.Wright of RAE, Mr.Barry Weeks of British Airways and Mr.K.W.Allen of The City University in the preparation of this paper is gratefully appreciated.

REFERENCES

1. The design of bonded structure repairs.
   K.B.Armstrong
   International Journal of Adhesion and Adhesives, January 1983 pp37-52

2. Durability of structural adhesives.
   Edited A.J.Kinloch
   Applied Science Publishers Ltd 1983

3. Transport of water in epoxide adhesives - Ph.D Thesis.
   Janet L. Tegg
   Leicester Polytechnic

4. Effects of curing on the glass transition temperature and moisture
   absorption of a neat epoxy resin.
   Ned D. Danieley
   Journal of Polymer Science, Polymer Chemistry Edition, Vol.19
   pp2443-2449 (1981)

5. The effect of water on the physical properties of epoxides — Ph.D
   Thesis.
   R.J.A.Shalash
   Leicester Polytechnic

6. Moisture absorption and desorption of composite materials.
   Chi-Hung Shen and George S. Springer
   Journal of Composite Materials Vol.10 (January 1976) pp2-20

7. Structural adhesive joints in engineering.
   Robert D. Adams and William C. Wake
   Elsevier Applied Science Publishers 1984

8. A comparison of acrylic adhesives for bonding aluminium alloys after
   using various surface preparation methods.
   K.W.Allen, T.Hatzinikolaou and K.B.Armstrong
   International Journal of Adhesion and Adhesives Vol.4, No.3, July
   1984 pp133-136

9. A comparison of different grades of an acrylic adhesive for bonding
   an aluminium alloy.
   K.W.Allen, L.Greenwood and K.B.Armstrong
   International Journal of Adhesion and Adhesives, Vol.5, No.3, July
   1985 pp149-151

10. Technology of carbon and graphite fibre composites.
    John Delmonte
    Van Nostrand Reinhold 1981

11. Effect of carriers on the environmental stability of adhesive joints.
    D.M.Brewis, J.Comyn, B.C.Cope and A.C.Moloney
    Final Report September 1979 Leicester Polytechnic School of Chemistry

12. Matrix deformation and fracture in graphite-reinforced epoxies.
     Walter L.Bradley and Ronald N.Cohen
     A.S.T.M. STP 876 Delamination & Debonding of Materials pp389-410

13. A review of the influence of absorbed moisture on the properties of
     composite materials based on epoxy resins.
     W.W.Wright
     R.A.E. Technical Memo MAT 324

14. Adhesives for bonding graphite/glass composites.
     Probir K.Guha and Joseph N.Epel
     Adhesives Age June 1979 pp31-34

15. Resin properties Part "A" cast resins.
     D.B.S.Berry, B.I.Buck, A.Cornwell and L.N.Phillips
     Prepared and published by Yarsley Testing Laboratories, The Street,
     Ashstead, Surrey under M.O.D. Contract No. K/LT32B/1932, October
     1975.

16. The effect of observed climatic conditions on the moisture
     equilibrium level of fibre-reinforced plastics
     T.A.Collings
     Composites Vol.17, No.1, January 1986.

17. A comparison of fracture toughness of matrix controlled failure
     modes: delamination and transverse cracking.
     Shaw Ming Lee
     Journal of Composite Materials Vol 20,March 1986

18. Failure mechanism of delamination fracture.
     Shaw Ming Lee
     8th Symposium on Composite Materials Testing and Design. 29th April
     - 1st May 1986, Charleston South Carolina.

19. Fracture of adhesive joints and laminated composites.
     Shaw Ming Lee
     International Symposium on Non-Linear Deformation,Fracture and
     Fatigue of Polymeric Materials,Chicago,Sept 8-13,1985.

20. Correlation between resin material variables and transverse cracking
     in composites.
     Shaw Ming Lee

21. Factors affecting the compression strength of aligned fibre
     composites.
     M.R.Piggott and B.Harris
     Proceedings of the Third International Conference on Composite
     Materials, I.C.C.M.3,Paris 1980
     Journal of Materials Science 19 (1984) pp2278-2288.

22. The characterisation of Mode 1 delamination failure in non-woven multi- directional laminates.
H.Chai
Composites Vol.15,No.4,Oct 1984.

23. The effect of resin failure strain on the tensile properties of glass-fibre reinforced polyester cross-ply laminates.
K.W.Garrett and J.E.Bailey

24. Effect of temperature on the adhesive fracture behaviour of an elastomer epoxy resin.
Willard.D.Bascom and Robert.L.Cottington
Journal of Adhesion,1976,(Vol.7) pp333-346
Gordon and Breach Science Publishers Ltd

25. Moisture sorption and desorption in epoxy resin matrix composites.
Charles.D.Shirrell
23rd National S.A.M.P.E. Symposium, Selective Application of Materials for Products and Energy, Disneyland Hotel, Anaheim, California,May 2-4,1978 pp175-191

26. Use of new epoxy resin systems for wet filament wound ,high performance structures.
Ralph.W.Hewitt,Laurie.M.Schlaudt,David.C.Bonner.
31st International S.A.M.P.E. Symposium April 7-10, 1986

27. The mechanics of bonded joints.
R.D.Adams
Paper C 180/86,International Conference, Structural Adhesives in Engineering, University of Bristol,June 1986.
Published Proceedings of the Institution of Mechanical Engineers.

28. Stress analysis concepts for the adhesive bonding of aircraft primary structure.
R.B.Kreiger
Paper C 174/86 International Conference, Structural Adhesives in Engineering, University of Bristol.
Published Proceedings of the Institution of Mechanical Engineers.

29. Operational durability of civil airframe structures.
R.J.Schliekelmann, Fokker B.V.
15th Annual Conference, International Federation of Airworthiness, Amsterdam 4-6 Nov. 1985.

174

30. A novel,damage tolerant,toughened epoxy resin.
    Charles.A.Swartz
    31st International S.A.M.P.E. Symposium April 7-10, 1986 pp163-176.

31. A new high performance epoxy resin for advanced composites.
    M.Chaudhari.
    31st International S.A.M.P.E. Symposium, April 7-10, 1986 pp563-570

32. Resin-hardener systems for resin transfer moulding.
    W.D.White
    31stInternational S.A.M.P.E. Symposium April 7-10, 1986, pp622-634.

33. Adhesive bonding of contaminated carbon- fibre composites.
    B.M.Parker
    Paper C.164/86. International Conference .Structural Adhesives in
    Engineering. University of Bristol, June 1986.
    Published Proceedings of the Institution of Mechanical Engineers.

Chapter 9

SURFACE INTERACTIONS BETWEEN METAL AND RUBBER

K. AB-MALEK and A. STEVENSON

Malaysian Rubber Producers' Research Association
Tun Abdul Razak Laboratory, Brickendonbury,
Hertford SG13 8NL, England.

1 INTRODUCTION

Interesting adhesion phenomena have been discovered during an experimental study of the puncture of rubber by cylindrical metal indentors. Puncture of rubber has been found to lead to the adhesion of a thin rubbery layer on the metal surface. In some cases this has been associated with an unexpected wear rate of metal by rubber.

There are a number of practical applications where surface inter-actions between rubber and more rigid materials are important. Tyres in contact with concrete roads and sealing rings in contact with steel surfaces are examples. In both cases the interactions can involve the wearing of the harder more rigid material by contact with softer rubber. The fact that this can occur suggests that the surface mechanisms involved are not entirely obvious. Some recent work has suggested that polymer radicals formed by the rupture or degradation of a polymer surface can react chemically with metal oxide layers and act either to decrease or increase rates of wear in different circumstances. Thus Vinagradov et al [1] found in the course of investigating the sliding of polymer discs against metal surfaces that modification of the contacting surfaces could occur both due to polymer degradation (associated with local frictional heating) and also due to metallization of the polymer surface. Gorokhovskii et al [2] also reported modifications to the structure of steel chips ground together with polymethyl methacrylate (PMMA) associated with

progressive mechanical degradation of the polymer. Gorokhovskii et al
also found that the addition of a small amount of polymer (5%) to
abrasive particles could increase the rate of wear of metals by a factor
of 2-3. Gent & Pulford [3,4] have also reported chemical effects. The
wear rate of steel razor blades held against rotating solid rubber wheels
was for example reported to be almost 70 times greater for polyisobutylene-
co-isoprene than for ethylene propylene rubber of similar hardness. The
authors attributed this effect to the direct attack upon metals of free
radical species generated by the mechanical rupture of elastomer molecules.

In the present study surface interaction associated with the
adhesion and wear of cylindrical metal indentors used to puncture surfaces
of solid rubber blocks have been investigated. The condition of the metal
indentor after rupture has been investigated both by electron microscopy
and by use of contact angle measurements which can provide [5] a
particularly sensitive indication of changes in the metal surface. Some
changes are reported and their correlation with both changes in 'puncture
loads' and with metal wear rates are explored. In the light of this
evidence there is some discussion of the likely role of polymer radical
interactions.

2 EXPERIMENTAL METHOD

The basic puncture test consists simply in forcing a metal cylinder
(the indentor) into a rubber block until the surface ruptures. The load
at rupture is called the 'puncture load' and is recorded as the peak on a
load/deformation plot. Each puncture test was performed on an Instron
testing machine and successive punctures always involved the rupture of
new material. The rubber blocks used were both of sufficient area to
allow a large number of successive adjacent punctures without mutual
interference and of sufficient thickness to eliminate any thickness
dependence in the results. The indentors were of 1.5mm diameter and about
4cm in length, the minimum required to accommodate the large surface
deformation that occurs for soft rubber prior to rupture. Most of the
puncture results reported involved the same indentation rate of 0.5cm/min
although the effect of rate was investigated for the range 0.5-20cm/min.
The indentor ends were polished to a mirror finish using a Kent lapping
machine and various grades of grinding paste. The aim was to produce as
sharp 90° corners as possible and subsequent examination of the indentor
corners with a scanning electron microscope (SEM) showed that corner radii

177

in the range 1-5μm had been achieved. The harder metals produced the smallest initial corner radii. Figure 1 shows a typical initial 'corner' for tungsten carbide.

Fig.1 Typical initial indentor corner for tungsten carbide.

Unless otherwise stated, the punctures were carried out in clean dry conditions with no external lubrication. The type of rubber used initially (vulcanizate A) was natural rubber crosslinked (or vulcanized) with 2.5 parts of Sulphur per hundred of rubber and containing the usual ingredients for a 'conventional' vulcanizate used in engineering applications, but without carbon black. This produces a vulcanized network with polysulphidic crosslinks. Other cure systems with less sulphur (monosulphidic crosslinks) and with dicuml peroxide (carbon-carbon crosslinks) were also investigated. Full details of the vulcanizate formulations are given in Table 1.

A typical force/deformation plot, obtained as the cylindrical indentor is forced into the rubber surface, is shown in fig. 2. At the point of puncture (or rupture) there is a sudden drop in load, and as the indentor is withdrawn some force in the opposite direction occurs due to rubber/metal friction. The recorded value of puncture load could be highly reproducible for given conditions of indentor and rubber vulcanizate type. Vulcanizate A, for example, when punctured with a sharp cornered 1.5mm titanium indentor showed a coefficient of variation of only 2% after the initial 10 events.

3 EFFECT OF REPEATED RUBBER/METAL CONTACT ON PUNCTURE LOAD

Over the first few punctures an unexpected reduction in puncture load was observed (see fig.3) which was qualitatively similar to that observed after the lubrication of the indentor by silicone oil.

TABLE 1:   Vulcanizate formulations

	A	B	C	D	E	F	G	H
Natural Rubber, SMR CV	100	100	100	100	100	100	100	100
Zinc Oxide	3.5	3.5	3.5	3.5	3.5	-	-	3.5
Stearic Acid	2	2	2	-	-	-	-	2
Zinc-2-ethylhexanoate	-	-	-	2	2	-	-	-
MOR[1]	-	-	-	1.7	1.7	-	-	-
TBTD[2]	-	-	-	0.7	0.7	-	-	-
Carbon black, SRF (ASTM-N-672)	-	-	20	-	50	-	-	85
Process Oil[3]	-	-	20	-	-	-	-	5
Antioxidant (Nonox ZA)[4]	-	-	-	2	2	-	-	-
Antioxidant/antiozonant, HPPD[5]	3	3	3	-	-	-	-	3
Antiozonant Wax[6]	2	2	2	-	-	-	-	2
CBS[7]	0.7	0.7	0.7	-	-	-	-	0.7
Sulphur	2.5	4	2.5	0.7	0.7	-	-	2.5
Dicumyl peroxide	-	-	-	-	-	0.5	5	-

Cure schedule:

	A	B	C	D	E	F	G	H
Temperature °C	141	141	141	140	140	140	140	141
Time hours	1	1	1	1	1	2	2	1

1   N-oxydiethylenebenzothiazole-2-sulphenamide
2   Tetrabutylthiuram disulphide
3   Fina Process Oil 2059 (Petrofina)
4   N-Isopropyl-N'-pheny-p-phenylenediamine
5   N-(1,3-dimethylbutyl)-N'-phenyl-p-phenylenediamine
6   Sunproof Improved (Uniroyal)
7   N-cyclohexylbenzothiazole-2-sulphenamide

179

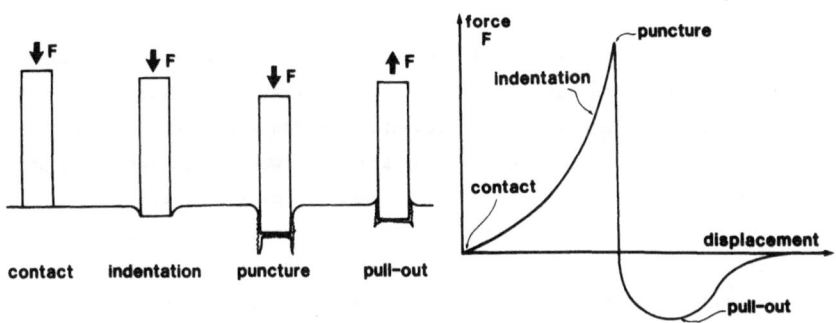

Fig.2(a) Schematic view of puncture   Fig.2(b) Typical force/deformation
          test stages.   plot for puncture test.

Initially, the indentor was fabricated from commercially pure titanium (99% Ti). However, repeating the experiment with indentors made from several different metals, namely a titanium alloy (Ti-89%, Al-6%, V-4%), tool sheet (Fe-70%, C-0.75%, W-18%, Cr-4.7%, Co-5.5% V-1.5%, Mo-0.4% Si-0.2%, Mn-0.2%) silver steel (Fe-99.25%, C-0.75%) and tungsten carbide produced the same phenomenon. It was found however that if the indentor was recleaned with acetone between each successive indentation, the phenomenon did not occur and the puncture load remained at the initial higher level.

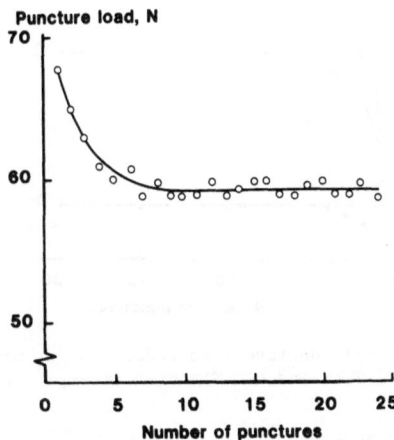

Fig.3 Reduction in puncture load over the first few events for a
       vulcanizate A. Titanium indentor at 0.5cm min[-1].

These observations suggest the adhesion of a layer of lubricating rubbery material on the metal surface immediately after puncture which then acts to reduce the puncture load. Scanning electron microscopy failed however to reveal any difference in appearance between clean indentors and those after several punctures. Chemical analysis of the constituents of the solvent wash used to clean the indentors also failed to detect any material clearly attributable to vulcanized rubber. This result suggests that any surface layer must be extremely thin, i.e. less than about 10nm.

The experiment was also repeated with a series of different vulcanizates (A-D) cured with different amount of sulphur. The results (shown in fig.4) demonstrate that increasing the level of sulphur to 4pphr (vulcanizate B) increased the effect whereas reducing it to 0.7pphr (vulcanizate D) eliminated it altogether. Incorporating 20pphr of carbon black into a vulcanizate (C) otherwise identical to vulcanizate A had the effect of substantially reducing the extent of the puncture load decrease. It was however still present.

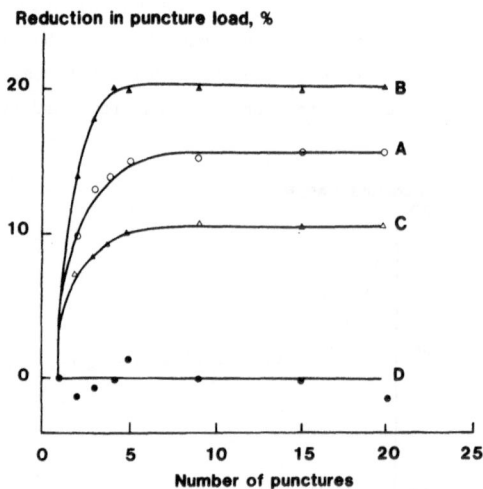

Fig.4 Percentage initial puncture load reduction for natural rubber vulcanizates A, B, C and D. Titanium indentor at 0.5cm min^{-1}.

4 CONTACT ANGLE MEASUREMENTS

As a more sensitive means of detecting the presence of a thin adhering rubbery layer on the metal indentors, contact angle experiments were performed. A metal indentor was immersed in distilled water and then

carefully raised through the liquid surface. The contact region between
the water and the metal was enlarged by projecting its silhouette onto a
translucent screen, as shown in fig.5. The screen was marked with angular
graduations, which enabled the angle of contact to be measured with
reasonable accuracy. The reliability of this technique was demonstrated
by using it to measure the contact angles of nylon and polytetrafluoroethylene (PTFE) rods. These were found to be 10° and 104° respectively,
in reasonably good agreement with literature values of 0° and 108° [6]
obtained by conventional contact angle tests on flat plates. The
technique adopted here effectively measures retarding contact angles, which
were found to be more reproducible than advancing angles. Since each of
the metals tested had shown the same puncture load reduction, most of these
experiments were with titanium indentors. This avoided the complication
(with the steels) of forming corrosion products in water and dramatically
changing the contact angle. Confirmatory experiments showed that provided
the latter was avoided, the choice of metal was irrelevant.

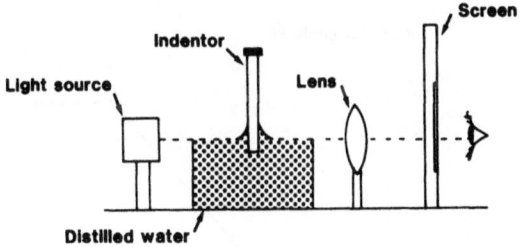

Fig.5 Experimental arrangement for contact angle measurements.

Vulcanizate A showed a substantial and progressive increase in
contact angle from 10° up to an 'equilibrium' value of about 75°. This
increase was quite distinct and was well outside experimental error.
Experiments with the other vulcanizates showed the phenomenon to varying
degrees as shown in fig.6. Contact angle increase was found to correlate
well with puncture load decrease, as illustrated by fig.7. This provides
evidence for the adhesion of a lubricating layer of rubbery material on the
indentor surfaces. Separate experiments with metal rods coated with rubber
latex yielded contact angles of about 80°. Drops of the same distilled
water on flat moulded surfaces of vulcanizates A-D formed contact angles
consistent with those shown in figs.6 and 7 as the equilibrium values
for indentors used to puncture the same vulcanizates.

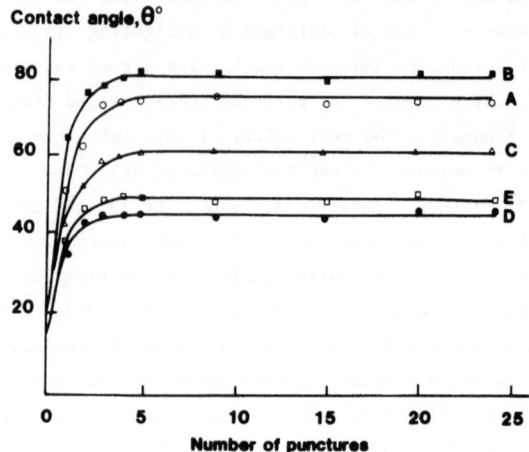

Fig.6 Contact angle increases for initial puncture events for
vulcanizates A, B, C, D and E.   Indentation rate 0.5cm min^{-1},
titanium indentor.

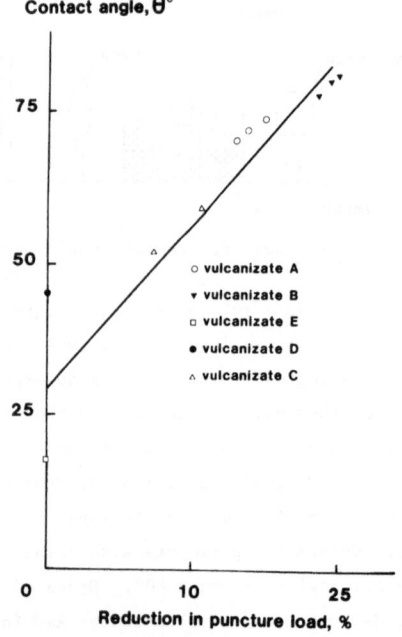

Fig.7 Reduction in puncture load plotted against contact angle for
vulcanizates A, B, C, D and E.

In the experiments described so far the duration of contact between the indentor and ruptured rubber surfaces was approximately 3 minutes. A series of experiments with vulcanizates B showed that reducing this 'dwell time' by increasing the rate of indentation from 0.5 to 5.0 cm/min significantly reduced the resultant contact angle from 80° to 50°, as shown in fig.8. Thus time of contact between rubber and metal may be an important factor. It was found subsequently that if the indentor was not withdrawn after the first indentation but was left in contact with the freshly ruptured rubber for periods of time of up to 7 days, the contact angles were even higher, approaching 95°, (see fig.8). In a control experiment the same total rubber/metal contact times were accumulated by means of repeated indentations at 5cm/min. This however produced no further increase in the effect beyond that recorded for the first five indentations. This suggests that some oxidation of the rubbery layer may occur during or immediately after its formation, producing a product with a lower contact angle in water.

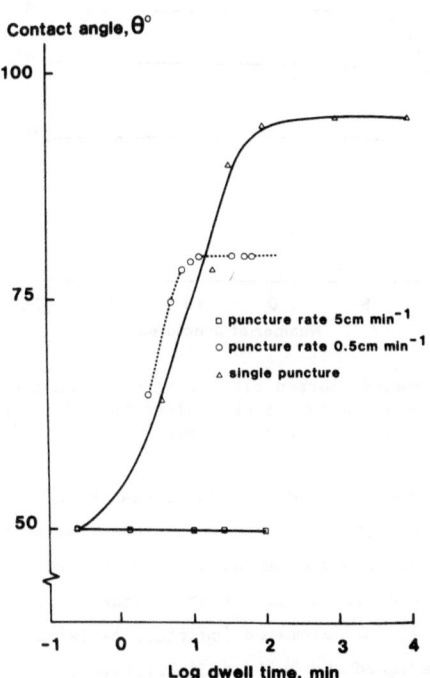

Fig.8 Effect of rubber/metal dwell time on contact angle using vulcanizate B and a titanium indentor.

Contact angle measurements appear much more sensitive to the type of surface changes under discussion than puncture load reductions - when in some cases the expected reduction would be within experimental error. An example of this can be seen in the effect of incorporating carbon black. A clear progressive reduction in the contact angle effect was observed for a series of vulcanizates (A, C & H) with the same sulphur cure system but with increasing amounts (0, 20pph and 85pph) of carbon black (SRF type) respectively. The results are shown in fig.9. At the same time, increasing carbon black content also increases the scatter in punture load measurements so that it is most difficult to determine the corresponding puncture load reductions unequivocally. This result suggests that the presence of carbon black may act to inhibit the adhesion of the rubbery layer or to remove it once formed.

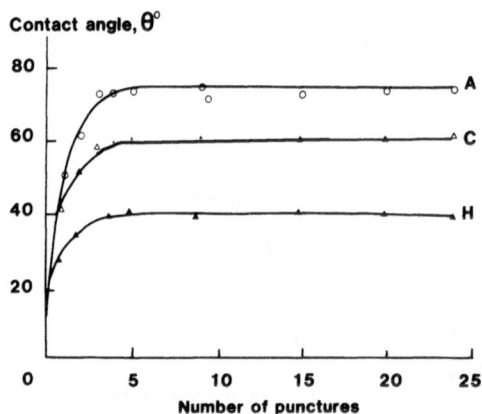

Fig.9 Effect of increasing carbon black content on contact angle effect. Vulcanizate A - no carbon black, vulcanizate C - 20pphr carbon black and vulcanizate H - 85pphr carbon black.

5 EFFECT OF INTERFACIAL CONDITIONS ON METAL WEAR RATES

The final stage of this investigation investigated the effect of the surface interactions described on metal wear rates. For this purpose, metal indentors were used to puncture rubber blocks of the same vulcanizate several thousand times. An automated 'puncture table' controlled by a microprocessor was designed and built to facilitate the accumulation of such large numbers of punctures. This automatically moved the rubber block to a fresh position while the test machine crosshead was raised ensuring that puncture always occurred on fresh rubber at a prescribed minimum distance from any previous puncture. The puncture rate was increased to 20cm/min for these experiments.

Examination of the indentor tip on an electron microscope enabled the change in corner radius to be measured and the volume of metal removed by puncturing to therefore be deduced. Figure 10 shows a typical series of SEM photographs for tool steel indentors. Photographs were taken at several points around the indentor circumference and average values taken for the change in corner radius. This technique proved to be reasonably accurate in detecting changes in corner radius of about 5μm and above. These radius changes were used to determine metal wear rates in terms of metal volume removed (μm^3) per puncture. The results are summarised in Table 2 for various different metals and different rubber vulcanizates.

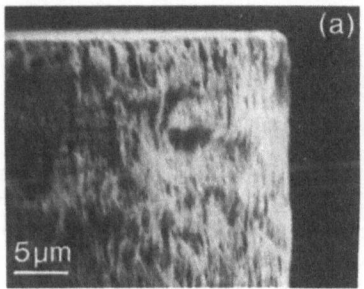 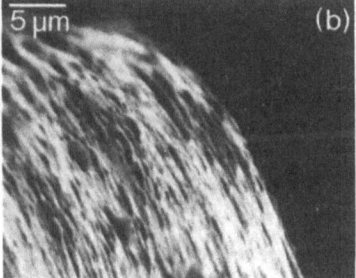

Fig.10 SEM photographs of the corners of tool steel indentors.
(a) - before   (b) - after 10^4 puncture

The type of metal used for the indentor is clearly the dominant factor in determining wear rates. There is a general trend for the hardest metals to exhibit the lowest wear rates, and, as shown by fig.11 in most cases wear rate correlates with metal hardness. There is an interesting exception to this in comparing the results for titanium with the results for silver steel. Titanium consistently showed very much lower wear rates than harder silver steel indentors. Titanium showed wear rates below the general trend while silver steel showed rates sharply above it. This suggests that metal hardness is not the only factor.

TABLE 2:   Indentor wear rates

Vulcanizate	Vulcanizate Hardness (IRHD)	Indentor	Indentor Hardness kg/mm^2	Wear Rate (μm^3 per puncture)
E	56	SS	300	1000 $\pm$ 100
E	56	Ti-2	152	300 $\pm$ 20
E	56	Ti-64	350	204 $\pm$ 14
E	56	HSTS	1200	104 $\pm$ 10
E	56	TC	1300	7 $\pm$ 1
A	45	Ti-64	350	96 $\pm$ 8
A	45	SS	300	34 $\pm$ 4
A	45	HSTS	1200	10 $\pm$ 2
D	32	Ti-2	152	80 $\pm$ 9
D	32	SS	300	91 $\pm$ 10
D	32	HSTS	1200	13 $\pm$ 2
D	32	Ti-64	350	120 $\pm$ 10
B	47	SS	300	8 $\pm$ 2

SS   - silver steel
Ti-2 - commercially pure titanium
Ti-64 - titanium alloy
HSTS - tool steel
TC   - tungsten carbide

Rubber vulcanizates containing carbon black filler gave significantly higher wear rates than the same rubber formulation unfilled. Indeed the fine 'scoring' marks which could be observed on all indentor surfaces were more pronounced for cases where carbon black filler had been present. The size of these score marks was about 500nm, which is somewhat greater than the known carbon black particle size of 70nm.

187

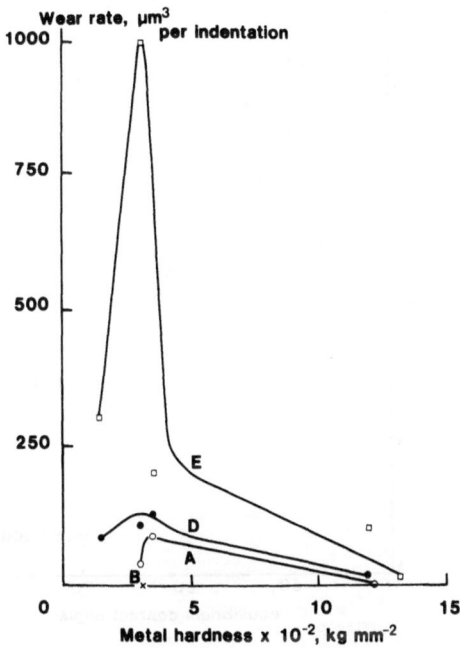

Fig.11 Metal wear rate plotted against metal hardness for
vulcanizates A, B, D and E.

The results of experiments with the 3 unfilled sulphur-cured
vulcanizates A, B and D are shown in fig.12 by plotting wear rates against
'equilibrium' contact angle for each vulcanizate. There is an unexpected
trend showing that the hardest rubbers produce lower wear rates. This is
in fact precisely the trend of increasing self-lubrication indicated by
contact angle measurements. In addition, the tear strengths of vulcanizates
B and A are higher, which might be expected to cause higher wear rates by
producing a greater strain energy at the tip of the indentor. However with
a titanium indentor there was a 50% greater wear rate in vulcanizate D than
in A. The tear strength of vulcanizate D was lower than than of vulcanizate
A. This suggests that the rubbery layer detected by the constant angle
measurements lubricates the indentor and inhibits wear at its corner.

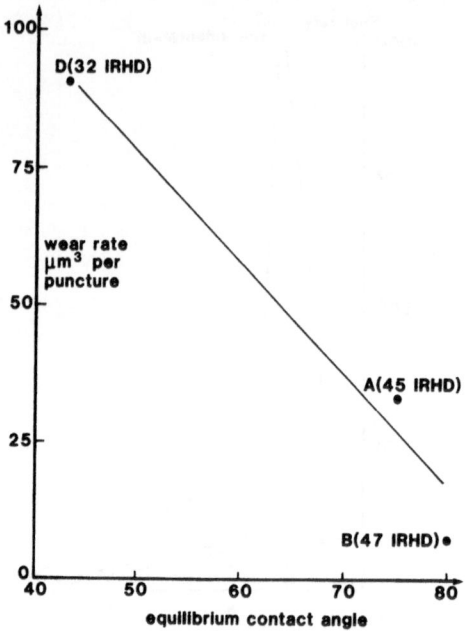

Fig.12   Wear rate of silver steel plotted against equilibrium contact
         angle of vulcanizates A, B and D.

Some additional wear rate experiments were carried out in the presence
of an applied liquid lubricant (TRIFLON - a PTFE based fluid) with three
different metal indentors, (titanium, tool steel and tungsten carbide) in
vulcanizate E. The wear rates were indistinguishable for the titanium
indentor but reduced by the presence of the lubricant for tool steel and
tungsten carbide by about a factor of 2. These results are shown in
Table 3. By way of comparison, the tool steel wear rate was a factor of
10 lower in vulcanizate A without addition of a liquid lubricant. The
addition of liquid lubricants would therefore appear to have less effect on
metal wear rates than the type of rubber vulcanizate used. The factors
controlling the metal wear rates thus appears in most cases to be, in order
of importance, first the metal hardness, second the amound of carbon black
incorporated in the rubber and third lubrication effects at the metal/rubber
contact zone - the self lubricating rubbery boundary layer being somewhat
more effective than added liquid lubricants.

189

TABLE 3:   Indentor wear rates on vulcanizate E lubricated and unlubricated

Test condition	Indentor	Hardness kg/mm^2	Wear rate (μm^3 per puncture)
Unlubricated	Ti-64	350	196 $\pm$ 15
Lubricated	Ti-64	350	208 $\pm$ 12
Unlubricated	HSTS	1200	104 $\pm$ 8
Lubricated	HSTS	1200	40 $\pm$ 5
Unlubricated	TC	1300	7 $\pm$ 1
Lubricated	TC	1300	3 $\pm$ 1

6 DISCUSSION

The present investigation has explored some of the surface interactions involved in the puncture of rubber by metal cylinders. The fact that soft unfilled rubber blocks can cause progressive metal wear itself suggests that the mechanisms involved may not all be obvious. Previous work has compared wear rates of different polymers and found the results to depend strongly on the type of polymer degradation caused by rupture and on the likely reaction of degraded polymer radicals with the metal oxide layers [4]. In the present case attention was restricted to one polymer type and other aspects of the rubber vulcanizate explored, namely the effect of the crosslinking system and the presence of carbon black filler.

Omitting the antioxidants and antiozonants from vulcanizate type A did not alter the phenomenon, nor did substituting the stearic acid for zinc-2-ethylhexanoate. This traces the origin of the rubbery layer to the nature of the crosslinked rubber network. Experiments with a lightly crosslinked peroxide cured natural rubber (vulcanizate F) did show a small effect, especially in contact angle changes, but this was not increased by increasing the crosslink density tenfold (vulcanizate G).

The results considered together suggest the following picture of what happens at the indentor tip.

The rubber surface first becomes elastically deformed as the metal indentor is forced into it. There may be some local slip at the edge of the indentor surface in contact with the rubber. The extent of such slip will be influenced by any surface lubrication and will have an effect on

the force required for the indentor to rupture the rubber surface (the
puncture load). The more lubrication there is the more the local slip and
the lower the puncture load. When the rubber ruptures, segments of the
molecular chain break and some may be released as free radicals. This
occurs while the metal oxide surface of the indentor is in intimate contact
with the rubber. Under these conditions the stability of the free radicals
formed may determine subsequent reactions to the metal oxide layer.

It is well known that clean freshly moulded rubber can exhibit
substantial adhesion to a clean glass surface [7] and that the energy of
adhesion increases with the time of contact, as dwell time. The strength
of adhesion of molecular segments of the freshly ruptured rubber may adhere
similarly under the action of 'Van der Waals' secondary intermolecular
forces. The strength of this adhesion may increase in time due to increased
packing of molecular segments on the surface or due to some intrinsic
increase in the adhesion of individual segments. When the indentor is
withdrawn, some of the molecular segments remain on the metal leaving a
thin rubbery covering. This is illustrated schematically in fig.13. The
experimental results suggest that there is a progressive increase in the
extent of this covering with repeated punctures. A monomolecular layer is
sufficient to radically alter contact angles - it could therefore be that
the packing of ruptured rubbery molecular segments increases with time up
to a complete covering (see fig.13). The experiments showed a strong
increase in contact angle with dwell time, suggesting the latter to be
influential in the mechanism of rubbery layer adhesion.

Different types of rubber network may release different types of
molecular segment - or even segments of different length. The present
results showed a large difference between mono-sulphidic and poly-sulphidic
networks. The latter showed the larger increase in contact angle and the
lower wear rates - indicating the formation of a more highly lubricating
rubbery layer. It is known that polysulphidic crosslinks are good
scavengers of carbon free radicals and ensuing reactions will produce less
reactive free radicals [10]. Thus, in a polysulphidic network vulcanizate
(A and B) there will be less available active free radicals to react with
oxygen. Monosulphidic crosslinks are less reactive with the carbon free
radicals and therefore there would be more active free radicals to react
with oxygen and increase the oxidation of the rubber. Oxidized rubber is
expected to be more polar and would therefore exhibit a lower contact angle
as shown by the vulcanizate D which has a predominantly monosulphidic crosslink.

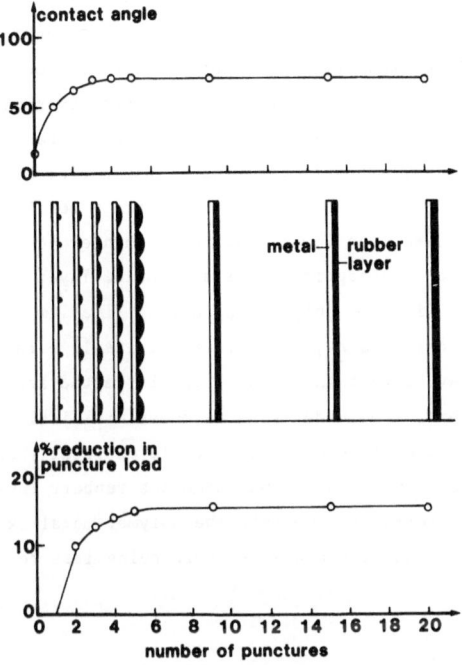

Fig.13 Effect of the extent of rubber covering on contact angle and
puncture load.

Once the rubbery layer is formed, there may be a second stage of
chemical reaction with the metal oxide surface. This may involve primary
chemical bonds and the production of a metal oxide-polymer complex which is
weaker than the metal oxide itself and detaches more easily from the surface.
Subsequent indentations may then partly remove this old layer and reform
with new rubbery material repeating the process. This provides a mechanism
for the wear of the harder metal by the softer rubber. In addition, the
softer and weaker vulcanizate D has more active free radicals than the
harder and stronger vulcanizate A and B, explaining why it shows the highest
wear rate.

Wear may also be caused by the fatigue of metal oxide asperities
under high local stresses during repeated punctures. If there is a
substantial amount of carbon black present the carbon black particles could
remove an old degraded layer more effectively than rubber alone. Their

presence in the rubber at rupture may also inhibit the formation of a complete layer. This would explain both the lower 'equilibrium' contact angle and the higher wear rate with carbon black filled rubbers.

The results showed a general trend towards increasing wear rate with decreasing metal hardness. A harder metal means in fact a higher local plastic yield stress. This fact dominated the differences in wear rate. There was however one anomaly in that the softest metal tested (commercially pure titanium) showed much lower wear rates than expected for its hardness. This may be due to an exceptionally stable oxide layer (TiO_2) which reacts much less readily with the polymer radicals. Titanium is noted for the stability of its oxide layer, the pure metal itself being extremely reactive in oxygen. It may be that the hardest metals tested are also noted for stable oxide layers. The underlying hardness will influence local deformations that could rupture the oxide layer locally and aid wear. The existence of a relatively thick and continuous rubbery layer although providing ample material to complete the polymer-metal oxide reaction may act as a lubricant and retard wear - there being less force to remove metal oxide products.

7 CONCLUSIONS

(i) During the puncture of rubber by metal adhesion of a thin rubbery layer can occur on the metal surface, causing a reduction in puncture load and an increase in contact angle values with a standard liquid.

(ii) Harder and stronger unfilled rubber vulcanizates can be associated with lower metal wear rates as can softer metals. Free radical formation and ensuing oxygen uptake are thought to be involved in those phenomena.

REFERENCES:

1. G.V. Vinogradov, V.A. Mustafaev and Yu. Podolsky. Wear 8 358-373 (1965).

2. G.A. Gorokhovskii, P.A.Chernenko and V.A.Smirnou. Sov. Mat. Sci 8, 557-00 (1972).

3. Idem, ibid. 10, 47-49 (1974).

4. Idem, ibid. 11, 32o-22 (1975).

5. A.N. Gent and C.T.R. Pulford. Wear, 49, 135-139 (1978).

6. A.N. Gent and C.T.R. Pulford. I. Mat. Sci 14, 1301-1307 (1979).

7. W.A. Zisman. Rev. Chem. Progr. 26, 13-51 (1965).

8. D. Tabor, "Friction & Lubrication of Solids", Part II. Oxofrd (1964).

9. G.J. Lake and A. Stevenson. Chapter in Adhesion 6 - ed. K.W. Allen, Applied Science (1980).

10. Poutsma M.L. in "Free Radicals" Vol.II, Ch. 14, edited by J.K. Kochi, (Wiley, 1973).